The Farm on Badger Creek

The Farm on
Badger Creek

❧

Memories of a Midwest Girlhood

Peggy Prilaman Marxen

Peggy Prilaman Marxen

WISCONSIN HISTORICAL SOCIETY PRESS

Published by the Wisconsin Historical Society Press
Publishers since 1855

The Wisconsin Historical Society helps people connect to the past by collecting, preserving, and sharing stories. Founded in 1846, the Society is one of the nation's finest historical institutions. *Join the Wisconsin Historical Society:* wisconsinhistory.org/membership

Cover and interior images are courtesy of Peggy Prilaman Marxen.

Printed in Canada
Cover designed by Sara DeHaan

25 24 23 22 21   1 2 3 4 5

Library of Congress Cataloging-in-Publication Data

Names: Marxen, Peggy Prilaman, 1947– author.
Title: The farm on Badger Creek : memories of a midwest girlhood / Peggy Prilaman Marxen
Description: [Madison] : Wisconsin Historical Society Press, [2021]
Identifiers: LCCN 2021001177 (print) | LCCN 2021001178 (e-book) | ISBN 978-0-87020-957-4 (paperback) | ISBN 978-0-87020-958-1 (epub)
Subjects: LCSH: Marxen, Peggy Prilaman, 1947– —Childhood and youth. | Girls—Wisconsin—Sawyer County—Biography. | Marxen, Peggy Prilaman, 1947– —Family. | Farm life—Wisconsin—Sawyer County. | Country life—Wisconsin—Sawyer County. | Farmers—Wisconsin—Sawyer County (Wis.)—Biography. | Rural families—Wisconsin—Sawyer County. | Sawyer County (Wis.)—Social life and customs—20th century. | Sawyer County (Wis.)—Rural conditions. | Sawyer County (Wis.)—Biography.
Classification: LCC F587.S29 M37 2021 (print) | LCC F587.S29 (e-book) | DDC 977.5/16043092 [B]—dc23
LC record available at https://lccn.loc.gov/2021001177
LC e-book record available at https://lccn.loc.gov/2021001178

For Sam and Shirley and Carol, who helped me remember.

And in remembrance of Richard, who would have loved the telling of these stories.

And for Twyla, Reed, Heather, Rebecca, Emma, and Nick: the next generation.

The Prilamans, 1952

Contents

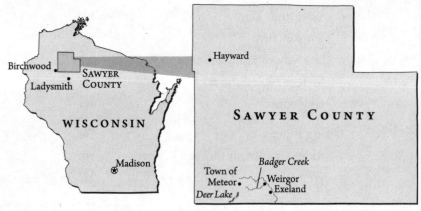

Sawyer County, Wisconsin

Prologue

Three farm families expected babies in the fall of 1947—the Hansons, the Halbergs, and the Prilamans. The fathers made a bet one Saturday night: the family of the last baby born would buy beer for a community party. Richard Halberg arrived near the end of September. Everyone thought the Hanson baby would be next, pretty much on time. As predicted, Verna Hanson appeared on October 7. Since I had not been due until November, my dad had been at a distinct disadvantage. However, showing up very unexpectedly, six weeks early on October 4, I left Verna's dad to spring for the party and my parents in a worried state after first driving twenty-seven miles in a rush to Dr. Pagel, in Ladysmith.

Fortunately, St. Mary's Hospital quickly admitted Mom. At 4 pounds 13 ounces, I required an incubator to breathe. Thirty-five days later, I reached a weight of 6 pounds 10 ounces and could breathe on my own. Mom and Dad brought me home on November 8, an early birthday present for my brother, Sam, who would turn five in five days.

My birth certificate read "Peggy Grace Prilaman": *Grace* after Mom's mother, my Grandma Walhovd; *Peggy*, not *Elizabeth*, as my Dad's mother, nicknamed Bessie, had expected. Peggy, after no one. Peggy—not even Margaret. Mom told me years later that Grandma Bessie had gotten on her high horse and said she'd anticipated a "far more dignified name." She had ended by pronouncing with a sniff, "'Peggy' is a *cow's* name."

This was fairly fitting, since I grew up on a subsistence dairy farm in a home nestled at the base of a hill alongside the burbling Badger Creek. As it carried water from the Blue Hills of northwestern Wisconsin, it met other creeks—Maple, Swan,

Badger Creek

Weirgor—before flowing to the Chippewa River and then to the Mississippi. I didn't think much about the creek as a child; everyone had a creek nearby, though none as close as ours, just outside our back-porch door. Visitors who mistakenly drove down our half-mile dead-end road were astonished at the peacefulness they found and by the proximity of our home to Badger's banks.

Though I claimed clear and sparkling Badger Creek as my own, I wasn't the first to set eyes upon its surging frigid waters, waters that long ago rushed down from glacial ridges, the way strewn with rocks and boulders. Long before I was born, the Lac Courte Oreilles Tribe had migrated to northern Wisconsin, trappers and traders had wandered the woods, a government surveyor had scrawled the name "Badger" in his notes (possibly naming the waterway after he'd watched a secretive badger dig its way to concealment), settlers had arrived, and then, beginning in the late 1800s, lumberjacks had stripped the area's pine forests.

Evidence of those times was as near as the old dirt logging road (known as the "tote road"), also just outside our back-porch door. It followed the curvature of glacially carved Badger Creek as it ran on its more or less gently sloping route that horses had been able to navigate—from Deer Lake a couple miles away, past our house, then toward the Chippewa and the Couderay Rivers.

Until the early 1900s, the tote road had carried "land lookers" and "timber cruisers" who estimated the value of each stick of standing timber, then crews of lumberjacks en route to the logging camp. The road was as uneven and rough as the huge pine logs once hauled over it by straining teams of horses. They had pulled sleighs over the ice-coated road, taking logs either to the sawmill at nearby Weirgor or for stacking along the Chippewa River, seven miles south, until, with melting snow in spring, lumberjacks plunged logs into the raging waters flowing to mills in Chippewa Falls.

By the time my farming grandparents settled in the area, the logging heyday was long over. During my childhood, the tote road functioned merely as a well-worn path. Along it, three close-knit homesteads dotted Badger Creek, less than two miles apart: ours; Uncle Jerry, Aunt Millie, and cousin Shirley's; and Grandma Bessie and Grandpa Sam's. Like their father before them, Dad and his brother, Jerry, eked out a living; hard as it was, they loved it. Our lives were not that much different from the 1800s-era settlers, even in the 1950s. Rabbits, partridge, deer, and trout provided food for our table. In the springtime, we boiled maple sap into rich amber syrup. We cut wood to heat our homes. Badger Creek's cold-water springs provided water not just to us but also to our horses, cows, and sheep that pastured in grassy openings.

Badger Creek and its companion tote road shaped our lives.

The creek fed us literally, mentally, and spiritually; we sat alongside its reflecting pools and pondered life's questions. Along the tote road we marveled at the wildflowers of spring, the vivid colors of fall, and the ever-deepening snows of winter. In the heart of the Blue Hills, specifically the Meteor Hills area that we called home, snow lies deep well into the spring. As Dad would say, the ice and snow kept us "locked tight."

In the springtime, the melting snow flowed away in rushing torrents down Badger Creek, washing away the icy snowpack of winter. In the summer, enormous claps of thunder rolled over the hills in hollow echoes and all-day downpours. The aphorism *If the creek don't rise* had real meaning to us. Badger Creek raged to heights, threatened flood, and lapped at our doorstep. It washed away the bridge on our dead-end road, temporarily stranding us, more times than I could count. In the warmth of summer, Badger could dry and slow to no more than a trickle before returning to soothing autumn's melodic flow. We tasted the bleakness of winter and the sweetness of summer; our lives were marked by the many moods of Badger Creek and unfolded beside Badger Creek, twisting like a stream.

As kids, Sam, cousin Shirley, and I built dams of rocks and stone, creating pools of water. We watched trout dart from side to side in overhanging banks, caught water striders, studied flora moistened by sparkling droplets, followed the creek's deer crossings and animal paths, and danced from rock to rock while exploring the tote road's creek-side boulders. Badger Creek and the tote road provided our 1950s entertainment and inspiration.

One day while at play, I unearthed what appeared to be a dirt-encrusted coin, a round metal piece wedged into the hard-packed ground. When Dad helped me scrub away the grime, I was puzzled by the words that became visible. Dad said the words were in French and that I hadn't found a coin but a watch fob, like the one Grandpa Sam attached to a leather lace and kept in his vest pocket. Of the many people who had walked it, I wondered, who had dropped that old fob on the tote road along Badger Creek?

Somehow, the fob vanished from my box of childhood treasures. Like memories of the past, connecting past and present, it had slipped away with little notice, growing more obscure with each passing year.

Today on Badger Creek, there are remnants of long ago: abandoned farmsteads that slump to the ground, smoke no longer curling from house chimneys; barns and hay sheds that have fallen into ruin; one-time gardens that remain unplowed; raspberry patches that have gone wild; overgrown apple orchards that stand forlorn; and saplings that have rooted in once well-trodden paths. However, a century has not extinguished the resilient growth of Grandma Bessie's pioneer-planted thicket of lilac bushes. These remaining signs tell of triumph and heartbreak, whispered voices, the sounds of cattle, and the melody of the old piano at twilight.

I think of home with the creak of harness leather, the whinny

Our log home on Badger Creek

of a horse, the call of a distant raven, the crack of an axe, the scent of soil in springtime, and new-mown hay in summer—and the aroma of coffee any time. The feel of flannel shirts and leather mittens. The sparkle of lantern light on a crust of snow.

If those pioneers were to return, we'd reminisce about the values instilled by people who connect us to northwestern Wisconsin and a 1950s rural lifestyle rapidly disappearing from memory.

I turn the car off of State Highway 48, slowing to a country crawl. Inhaling deeply, I stop at the bridge over Badger Creek. A long sigh escapes, and my shoulders drop in relaxation. I am home.

And I remember . . .

The Perfect Place

BEGINNINGS

In 1897, my grandfather Sam Prilaman promised his bride, "We'll settle when I've finally found the perfect place." Little did my grandmother realize that "settling" would take seventeen years.

Sam had exercised his itchy foot before ever encountering Bessie. After earning his teaching certificate from the Normal School in Valparaiso, Indiana, and teaching for a short time, he rode flatboats down the Wabash River, lit fireplaces in the women's dormitories at Purdue University, and, finally, worked as a farmhand all over northwestern Indiana. That's how he met Mary Elizabeth "Bessie" Rolls.

By the time Sam arrived on the scene, Bessie was nearly thirty years old, and her family consisted of her mother and siblings. Her homesteading father, Jethro, had died in 1894, leaving his wife, one surviving son, and six daughters to run the farm. Before emigrating from England, Bessie's well-educated mother had lived as the daughter of privileged English clergy. The rough living conditions in Indiana had shocked her; thus, she taught her family well—Bessie earned her teaching certificate, taught school, and gave piano lessons, earning enough to purchase a piano.

Sam and Bessie Prilaman's wedding day, Morocco, Indiana, 1897

After their marriage, Sam and Bessie farmed nearby for a few years. But not long after the birth of their first son, the sprawling prairie lands of Canada enticed Sam. In 1905, the three boarded a train bound for the tiny settlement of Starbuck, near Winnipeg, Manitoba. Their homestead ventures were fraught with hardship: relentless wind, freezing temperatures, and varied successes. Their last child, my father, Freeman Reed Prilaman, was born in Canada on Good Friday, April 9, 1909.

Nine years went by before the sugar beet crop failed and a prairie fire caused by sparks from a passing train destroyed their wheat crop. Sam decided to abandon his dream of establishing

Uncle Jerry and Dad in Canada

a homestead, auctioned his farm equipment and horses, and returned to Indiana in December 1913. Bessie had sent a vast collection of colorful postcards from Canada to her relatives over those years. They contain short messages about her loneliness and the joy she felt when she knew the jaunt to Canada would be over. Bessie was finally going home.

Yet by the end of January 1914, Sam was on the road again. He had purchased a new bicycle and boarded a train, leaving Bessie to care for their two sons, Gerald "Jerry," eleven, and Freeman, five, presumably with help from her home folks.

Sam rode the rails across the country: Cairo, Illinois; Missouri, Arkansas, Kentucky. He fetched up in Richmond and Roanoke, Virginia, where he found dozens of related Swiss-German Prilaman kin who had never moved from the ancestral home. Still he moved on: West Virginia; Washington DC; as far east as Forest Hill, Maryland; and as far west as Colorado.

At each stop, Sam explored the area by bicycle, talking with settlers and residents, assessing the potential for prosperity.

Occasionally he stopped to work; always, he posted a letter. Sam's letters back home commented on climate, growing conditions, soils, scenery, and the suitability of communities he visited. They also doled out advice to his long-suffering wife, as in the following from March 10, 1914, written from Forest Hill, Maryland:

> I know I've been gone for quite a while now; your complaints will not keep me from my search for the perfect place for set-tling; you must remain patient. I am not thin, but fat and in better health than when I started. You must get along and quit worrying. I must have time to get a place.

Finally, the news Bessie had been waiting for arrived. It is not known whether Bessie's growing impatience or Sam's own certainty spurred on his final choice. In 1916, Sam sent Bessie promotional postcards extolling the rich bounty of the lands in Sawyer County, Wisconsin: towering maples, oaks, elm, and other hardwoods; fine flowing streams; abundant game and large deer with rangy antlers; smiling gardeners in straw hats standing amid crops of peas, monstrous potatoes, and even tobacco; smiling children at a new school building; and healthy red poll cattle grazing freely. The postcards, however, were somewhat deceptive.

After all, once the lumber barons had stripped the stately white pines, land agents with the Upper Wisconsin Land Company colonized cutover lands by heavily promoting and selling "ready-made farms"—in reality, acres of rocky, pine-stump land. The land agents had stretched the truth of their claims, but Sawyer County *was* rich in resources. In addition, the rich-but-heavy clay soil held moisture longer and dried out slowly: good in a drought (but that would rarely be a problem). Sam wasn't alone in overlooking the short growing season; long, brutal winters; and rock-strewn ground beneath a bed of pine stumps.

Thus my grandfather became one of the first to purchase

land from the Upper Wisconsin Land Company, after which he returned to Indiana to fetch his small family. They loaded their possessions—shipping equipment, animals, and Bessie's piano—onto a boxcar, boarded a passenger train bound for Wisconsin, and waved goodbye to Indiana for good.

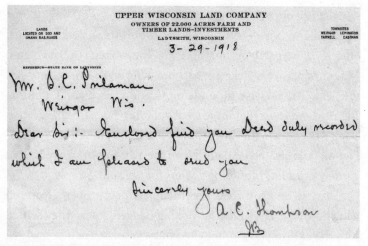

Original land purchase receipt of deed

The Prilaman family with a land agent and a relative, Exeland, 1916

Upon disembarking at the Exeland depot in the spring of 1916, Bessie and the boys, Jerry (then thirteen) and Freeman (seven), first set eyes on a land crisscrossed with old logging tote roads and dotted with defunct train trestles, tracks, abandoned cabins, and lumber camps; pine stumps were everywhere in this "perfect place."

The Prilaman family spent the daylight hours pulling stumps by hand and horse, cutting tall hardwood trees, picking rocks, and clearing a patch of land for crops. They weren't the only ones. Vast piles of rocks, picked in backbreaking work, lined every single field. The massive mounds on Grandpa Sam and Grandma Bessie's land attest that these were not folk who feared hard labor. Each day ended with exhaustion—but also with pride and satisfaction. The cleared land produced the ever-important vegetable garden and a few blooming flowers for the soul. Simple joys.

Situated on a tote road, the Prilamans' new property featured a log house on a hillside overlooking Badger Creek. They settled into the existing house and eked out a subsistence living with a team of horses, a cow or two, and chickens—just enough. During the Depression, they would build the home I knew, a white clapboard house constructed with wood from their forest.

The boys supplemented the family diet by hunting and fishing. Freeman, with his new friend, Ole, attended the newly built Valley View schoolhouse a mile away. Jerry, a rugged thirteen-year-old, headed straight to the woods. He joined one of the last pine logging crews during winter and spring, just in time to witness and take part in one of the final log drives in that part of the north country, floating logs down the Chippewa River.

Through the years, Badger Creek raced on. When Sam and Bessie's boys grew up, they too stayed in paradise; Badger Creek served as the family's connecting thread, like a fine watery filament.

COURTSHIP

Prilaman men seemed bent on establishing themselves on a piece of suitable land they deemed "perfect" for their future endeavors. Jerry married first and built a log home a stone's throw from his parents. Dad, however, exhibited a touch of his own father's wanderlust, venturing by train for a fall season and harvesting grain in the Dakota wheat fields, then looking for work in Wisconsin's largest city, Milwaukee; the latter cemented his decision to return home to work timber and the land.

Bessie wrote to her mother back in Indiana about him:

November 11, 1928
Dearest Little Mother,
Freeman came home on the 2nd of November dressed like a real gentleman. He did not get laid off and could have stayed all winter. I suppose that he was transferred and his wages would not be so much as what he could save to come home. So, he came in on the five o'clock limited and walked in on us just as we were getting up. He is a dear boy. Wish you could have seen him. Nice looking and fine built. He has bought forty acres of land west of us about a half mile, hired a man and is working up the timber into whatever it will make, the logs, wood etc. He had enough saved up to pay cash for it. He has Ole working for him. I hope he will make it, as it is his first venture.

All my love, Bessie

Dad, at nineteen, had purchased a "forty" (forty acres) just before the Depression. However, he hadn't been "transferred," as his mother supposed. Truth was, the tough working conditions imposed upon the older black men by their boss appalled him;

those men had dropped of exhaustion. Dad had seen enough of city life.

Elm, maple, oak, hickory, ash, and yellow birch covered Dad's forty acres. He took a team of horses, shouldered his axe, hired his best friend Ole, and went about the work of clearing for crops and pasture: cutting hardwood timber, digging out, and pulling or dynamiting pine stumps.

During the Depression, Dad continued to work his own timber as well as that of his neighbor, John Halberg. Halberg and his five sons had established a sawmill on Swan Creek, a mile away from the Prilaman homestead, housing most of their employees in the bunkhouse next to their rambling place. Lumberjacks like Dad worked in the woods or at the mill or both.

After a week spent hard at work, Saturday nights meant entertainment: that meant, in those days, piling into an old car and driving to a dance hall in nearby Exeland or any number of nearby communities. That's where Dad met Vera Walhovd, from Birchwood. Soon the two were spending Saturday nights or Sunday afternoons together whenever they could.

To the dismay of their parents and siblings, the dating went on for years. Seven, exactly. During their courtship, Dad lived with his parents and Vera with hers. Members of their families thought Freeman Prilaman and Vera Walhovd would *never* get married.

What took them so long? Hard times and lack of money. Like most farm folks, the Prilamans weathered the Depression better than most—relying on the produce and canning from their gardens and meat, eggs, and dairy from the farm. But Dad remembered one lean year, when in springtime there wasn't much left except for the withered cabbages and, after keeping some aside for planting, the last shriveled potatoes.

Most likely, Dad waited to marry until he had acquired enough land for a sustainable living. In 1936, he bought three more "forties"—one hundred and twenty acres. The fully wooded land on

Mom and Dad courting

Badger Creek cost only back taxes, which Dad paid at the Sawyer County Courthouse. Total cost: $10.52 (a steal even in 2019 terms: $179.95.)

Clues to their lengthy, long-distance—fifteen-miles-apart—courtship may also be found in snippets from Mom's saved letters, tied with a light blue ribbon. Dad's letters recorded his patience and perseverance in overcoming obstacles thrown at their progressing relationship, some of which Dad created himself—namely, every November during hunting season, during one particular April in 1931, when he accidentally chopped his foot with an axe:

I was down to the Dr.'s office Saturday and he pulled the
stitches out, but told me to be sure and keep off my foot. If
it does not go bad I think I can get on it sometime this week
if I "misunderstand" the Dr. slightly.

Mom sent homemade candy by mail to take Dad's mind off
his injury. After thanking her, his letter continues:

I am still laid up. I did not get blood poisoning, although I
think I would have, if we had not given the cut prompt attention when it commenced to look bad. The Doc was right.
I had to keep my foot up in front of me and that was all the
scenery I had for six days. The next time I feel like doing
anything foolish, I think I will see how high I can kick. Don't
know if I can get over by the end of the week, but I will, unless my foot takes a notion to go on another spree.

But most of the interferences preventing Saturday night or
Sunday visits came from weather, transportation, road conditions, and work, and often from a combination of these.

January 1933:

It was sure cold this morning. The thermometer said minus
45 degrees at daylight, so I sat by the airtight stove until noon.
When I went out this afternoon. I succeeded in frosting my
nose a trifle.

March 1937:

Well, I almost came to see you last Sunday, my intentions
were good anyway. I started the motor but got the wires
mixed up and had quite a bit of trouble before I found the
cause and got it running. When I was ready to go had two

flat tires which had to be thawed out and fixed. Then the air got full of mist and frost and coated the outside of the windshield so bad I had to stick my neck out the window to drive. I figured out how far I had to drive and how slow I was having to drive so I turned about and came back home. When I got home it was below zero, and I have been congratulating myself ever since.

April 1938:

The frost is going out of the roads and the mud ruts are certainly deep, so deep that the axle drags. Got stuck three times and came close to it many more. The roads seem to be getting worse instead of better.

Well, there is a lot I would like to say but I am a poor letter writer, and it don't look quite right on paper anyway so will close. Will see you next week, if I can get through.

With lots of love, Freeman

On an ordinary Sunday afternoon—May 23, 1938—Freeman's car purred cooperatively enough to pick up my mom in Birchwood, and in the fresh and fragrant scent of springtime, the two drove away to Shell Lake. They summoned the Methodist minister at the parsonage and his wife as an attending witness. Freeman, twenty-nine, and Vera, twenty-six, tied the marriage knot. Their marriage was not entirely unexpected, but my parents' elopement sure was.

STARTING A LIFE TOGETHER

The newlyweds lived with Dad's parents, taking two years to build their own home on Badger Creek. For months, Dad and Mom walked the tote road to their site a mile and a half or so upstream, or in country-speak, "three forties." Together they built

Mom and Dad's wedding day, 1938

a two-room log house with leftover elm logs because there was no market for selling them at the time. Dad was more experienced at driving and controlling the team of horses, but only he knew the dangerous maneuvers that would be needed to ease the logs into place. That left Mom, a "town girl," to drive the horses and hoist the logs. One of the horses, named Joe, was young and spirited: one wrong move from him, or Mom, could spell disaster. But horses, Mom, and Dad completed the treacherous task without mishap.

While clearing the land for the log house, Dad came upon an axe looking much like an Indian trade axe, bearing the stamp "USID" (for United States Indian Department; these tools were used as payment to members of the Lac Courte Oreilles Tribe

COUNTY OF SAWYER
STATE OF WISCONSIN

REDEMPTION RECEIPT № 19207

HAYWARD, WIS., *Oct 16* 19*37*

Received of *Freeman Prilaman*

Ten Dollars, and *52* Cents

for the redemption of the following described property sold for Taxes, Costs and Charges of Sale together with interest at the rate as fixed by law or resolution, from the date of sale. Sale of 19 *37* Taxes of 19 *36*

NO. OF CER-TIFICATE	DESCRIPTION	SEC. OR LOT	TOWN OR BLOCK	RANGE OR WARD	FACE OF CERTIFICATE	INTEREST	TOTAL
2061	S¼ - NE	23	37	8	5 01	15	5 26
63	NE - N¼	"	"	"	1 28	04	1 32
65	SE - N¼	"	"	"	3 83	11	3 94

			ADVERTISING FEES	
			REDEMPTION FEES	
			TOTAL	10 52

COUNTY TREASURER

$10.52 for 120 acres, 1937

between 1838 and 1854). He made a handle for it and tucked it away in the corner of his newly built woodshed. Along with the ancient axe, Dad housed his double- and single-bit axes, wedges, mauls, and hatchets (his everyday tools), as well as his logging tools: crosscut saws, Swede saws, bucksaws, and cant hooks. They hung on walls along with leather horse harnesses, double trees, single trees, whiffletrees, and logging chains.

To set up housekeeping, Dad and Mom made the trip to the mercantile store in Exeland and purchased basic supplies: coffee, tea, sugar, a few spices, and flour for making bread. With only the essential pieces of used furniture—a bedstead and frame; an old but well-made round oak table; and a heavy, cast-iron, well-seasoned cook stove—they were in business. For now, water would come from a spring in the creek. The land would provide the wood for fuel.

To house their team of horses and handful of cows, Dad built a small barn, flat-roofed and unpainted, with rough boards. Along with trout from Badger Creek, venison from the woods, and vegetables from the garden, the cows and a few chickens would provide what they needed.

When World War II broke out, the military exempted Dad from duty because he produced food for the country—luckily, for Mom gave birth to my brother, Sam, on November 13, 1942. The small farming operation was well underway, and so was what would in five years become our family of four. My premature birth had been alarming in the context of the peaceful farmstead; despite that, I grew and suffered no health complications.

As the youngest of Grandma Bessie's three grandchildren, I was so painfully shy that I rarely said a word, happy to let others speak for me when I could not hide behind Mom's skirts. My silence allowed me always to be the quiet observer; my aunts and uncles called me "the dark-eyed girl who stood back to watch."

Sam and Peggy, 1948

CONNECTION TO THE PAST

As I grew older, I realized that, in various ways, we are connected to pathways of the past. Dad lived with one foot in the past: the wool clothes he wore, his lumberjack turns of phrase, and his appreciation of our area's history all told me that my life was not far removed from what had been. The past crept into Dad's voice as he greeted work mates from his logging days, with nicknames earned in lumber camps—Red and Moose and Skinny. They'd reminisce over their own stories as well as stories told to them, handed down from one generation of loggers to the next; tales of when men built railroad trestles named High Bridge, Crooked Bridge, Long Bridge, and Floating Bridge—the one that crossed a swamp; tales of when the Chippewa River and Menominee Railway had been built in the 1870s—'jacks back in the day had called the C. R. & M. the "Crooked, Rough, and Muddy"; tales of Saturday night poker games at lumber camps named Camp 5, Poppyville, Polish Camp, and Buck's Lake Camp.

Whenever winter set in, just as Dad and his mates had done at camp, he donned "long-handled underwear," or long-johns (the one-piece, neck-to-ankle kind, complete with a buttoned-up drop seat in the back), colorful checked or plaid wool shirts, dense wool lumberjack jackets, bulky woolen pants and socks, and wool hats with ear flappers.

He used colorful camp language when he offered to "rustle up some grub" for a meal. After cutting off thick-skinned generous slices from a slab of bacon, Dad fried it up in an iron skillet and made corn meal mush, flapjacks, and hunks of fried bread. To wash it down, Dad followed his recipe for heavy-duty lumber camp coffee, remarking in his lumberjack speak: "Fill yer pot, any pot, three-fourths fulla cold water. Over 'n open flame, get the water boilin'. After a bit, dump in several scoopsa coffee. Don't matter what kind. Let 'er boil. Boil the hell outta it." Dad would then get the dime store mugs ready, continuing, "Let 'er set, add a splasha cold water. Let the grounds settle. Pour, add a dasha cream. And give some to the cat." (Me too, for I started drinking coffee at the tender age of nine.)

Dad made the past seem present, as did many others in my life, notably Grandpa Sam and Grandma Bessie. I knew Grandpa Sam as a reserved, contemplative, and well-spoken man, perhaps because of his staid and principled upbringing as the son of pacifist members of the Church of the Brethren. Sometimes called Dunkards, named thus after their ritual of baptism by immersion, church members dressed in unadorned black clothing, without buttons. My dad remembered his Indiana grandparents that way, having seen them only the one time. (Perhaps my grandfather's desire to roam was a reaction to his upbringing?)

As her mother before her, Grandma Bessie had persevered while retaining a proper English home. She set a suitable table, loaded it with garden-grown food, and served tea in bone china cups. She was horrified if company caught her on wash day with nothing but brown bread in the breadbox, finding it highly inappropriate to serve anything but homemade white bread to her guests. Bessie shared her gift of gab, hospitality, bit of music on the piano, and outspoken opinions with all who came by.

And plenty of neighbors did, as well as the family who lived so close to her and Grandpa Sam on the banks of Badger Creek.

At Home

A SENSE OF PLACE

My first memories are of our two-room log house, our porch over-looking the creek, and the path to the outhouse. That well-used uphill path wound past tall trees and several huge pine stumps, making it an interesting short walk—if one weren't in a rush. In the wintertime, Mom kept under the bed a white enamel "thunder mug" with a red-trimmed lid. We used it when answering the call of nature instead of making a very chilly walk at night. Still, Mom had to empty the pot in the morning.

Dad took advantage of the secluded hillside toilet as a place of relaxation while, as the euphemism goes, taking care of business. Surveying his farm and surrounding countryside through the partially opened door on the old outhouse, his view looked down toward our log house and Badger Creek threading beside it. From his vantage point, Dad scanned Jerry's adjoining land, one of our crop fields, and our well house, with its pump and milk-cooling tank, and the outside tank where we watered the cows. (Most people had water in the barn, but we had to take our cows to a water tank at our well.)

Our barn and barnyard were behind the outhouse, another short distance up the hillside. The barn stood in a wooded opening along with a shed for calves and horses, a tractor shed (Dad used both horsepower and a tractor for farm work), and haystacks or, later, a hay shed. Dad had partially cleared a section of the wooded land behind our barn to become one pasture for our cows to graze.

Long before I was born, Dad bought forty acres of adjacent land from Bill and Anna Mulkey. The former Mulkey property consisted of mixed pasture, small fields, Badger Creek and one meandering branch (for watering the cattle), and more wooded land beyond to the west. All were reached by a short walk along a ravine, on the tote road through the woods.

Dad continued to acquire forty-acre plots of timberland adjoining our land whenever they came up for sale and the price was right; eventually we owned three hundred and sixty acres of mostly timberland. Like his father before him, he considered himself a good judge of the value of a healthy, growing forest containing good stands of timber—providing monetary savings far better than dollars in a sock under the mattress or even currency in the bank. Evidently, that belief ran in force throughout the Prilaman family.

Once, wardens from the Wisconsin Conservation Department came to Jerry's front yard attempting to capture an orphaned cub that had climbed up a sapling, just outside the house. The bear seemed to have no interest in coming down, so the warden suggested cutting down the tree. "I'll be g**-damned if you'll be doin' that here," Jerry blustered, retorting that if *that* was what they had in mind, they could vacate his property immediately. Fortunately, the cub came down, preventing a brawl and sparing the tree. Jerry remained firm until his death: the trees around his domain would stay untouched. He even requested that he not be buried in the treeless Exeland Cemetery but instead be placed with Millie's family in the Stone Lake Evergreen Cemetery where he could lie in a grove of giant white pines.

Like Jerry and Grandpa Sam, Dad felt connected to his trees. Sometimes he harvested mature trees as a supplement to farm income when necessary, such as when he announced: "We need to replace the old car; guess I'll take out some loggin' this winter." He frequently proclaimed, "You'll never make any money hirin' someone else to take down yer trees. Ya gotta do the loggin' yerself."

Mostly Dad culled his timber with forest management in mind, taking out trees that would "never amount to anything," those that were damaged, unhealthy, or overcrowded. Removing the never-amount-to-anything trees served a double purpose, allowing for new growth and heating our home. After hauling logs with his team of horses, Dad used the giant, round saw that was attached with belts to the tractor to "buzz" the wood into stove-length pieces. Then he swung his double-bit axe until the wood was small enough for me to haul to our back porch outside the kitchen.

The open back porch overlooking the tote road and Badger Creek was so close to the creek that Dad had raised it up a few steps in case of high water. In all seasons, it held the milking equipment and cream cans. In winter, it sheltered the mountain of wood I'd hauled and stacked for use in both the cook stove and the heating stove and served as our place to brush off winter's snow, take off spring's mud-covered boots, hang up summer's rain gear, and remove every season's dirty barn clothes and boots before entering the house. Smokey, our long-haired, black Labrador–Irish Setter mix, slept on the porch in good weather, usually in the company of a pile of cats.

In good weather, our kitchen entryway from the porch remained open, and thus we were always aware of the creek and the amount of water flowing in it. Most seasons, Badger Creek provided a symphony of sound: in spring, rushing ice and melting snow; in early summer and fall, raging muddy water after a heavy rain. Often Badger Creek lulled me to sleep and woke me up with a roar during downpours. In winter, however, drifted high with snow and frozen nearly solid, Badger remained hushed, silent.

When Sam was little, Dad had built a bridge over Badger Creek to connect our driveway spur on the tote road to the half-mile dead-end road leading to County Highway C. Also, at that time, Badger's cold-water springs still provided drinking, cooking, and bathing water year-round. But as soon as he could afford it, Dad had a deep well dug. Sam remembered how the local well driller used a divining rod, known as a dousing stick. The forked stick, when held in the hands of someone with long experience, proclaimed the spot where water would be found. Our water came from some one hundred and ten feet below ground.

Perhaps because Dad had grown up using only water from a Badger Creek spring, he was very proud of his well. On hot summer days, I helped Mom get Dad a cold pitcher of water. She pumped, then acted as bartender, singing out as in the radio ad running at the time: "What'll ya have?"

"Pabst Blue Ribbon!" I'd lustily call back.

In the stifling heat of summer days, nothing soothed thirst like a drink of water pumped straight from the deep well, fully appreciated only when served with lips pressed against the well's cool tin cup.

Sam and I had need of that refreshing water, occupied as we were with farm chores and playing in our yard, the woods, the pasture, and around the barn and fields. Our parents had need of it too: Dad milked the herd of around twenty-five cows while Mom fed the calves, kept the garden, and maintained the household.

When I was very small, Mom raised chickens in the coop beside the creek. Peeping boxes arrived via train in spring, and sometimes Mom had to keep them in the warmth of the kitchen

Peggy and the chickens

stove for a few days until the weather warmed. Raising chicks ended when I was three or four. I liked to play with the young chickens, feeling their soft feathers on my cheek when I clutched them, but one day Mom had to rescue me from my frantic attempt to outrun a pair of ornery roosters chasing me all over the yard. They had their comeuppance; one-by-one I watched Mom place each on the woodshed chopping block. Thus, I've seen firsthand what it means to "run around like a chicken with its head cut off."

Entering the two-room house from the porch into the kitchen, our main living space, we had only the one doorway to pass through to reach the bedroom. We all slept in that room together—Mom and Dad in a bed, Sam on a little cot, and me in a crib until age five. Only then did we first get electricity.

Before that, my parents lit kerosene lamps and compressed gas mantle lamps. The mantle hissed and sputtered at first, then the light grew in brightness, bathing the kitchen in glowing warmth. The pungent odor of kerosene or white gas hung in the air before dissipating. Every now and then during the summer, beautiful brown cecropia moths or lime green Luna moths were drawn toward the light filtering through our screen door. At dusk, Dad lit and carried kerosene lanterns to illuminate the barn and milking machine.

Although it was cozy, neither Dad nor Mom was satisfied with our small log home. Like most of our neighbors, Dad had learned building construction out of necessity, self-taught with help and advice from others who'd done the same. When I was about five years old, Dad made improvements to our home, adding what he called a "front room"—our new living room. The porchlike room captured the morning sun with its long bank of windows facing east and expanded the size of our house by nearly half. Dad cut butternut trees from our land and planed and sanded the wood so the lovely blond woodwork trimmed the length of the room. He cut a doorway through the original logs, from the kitchen to the new room, and installed another heating stove for the larger area.

In Rice Lake, Mom happily purchased our first pieces of living room furniture: a brand-new green davenport and matching chair (our first luxuries). She also arranged to have her mother's treasured upright piano brought from Birchwood. Family logging friends packed it up and had it delivered on a winter's day. Mom and I went out into the bright sunshine to meet the truck parked next to the high snowbanks lining our driveway. I was so excited that I dove into one, so deeply burying myself that Mom had to pull me out.

Just as it had been for Grandma Walhovd, the piano became our treasure. On summer nights, while Sam and Dad finished the milking chores, after Mom had cleared the table and done the dishes, and she and I had fed the calves, she finally sat down to rest at the piano keys. At twilight, the melody of one parlor song's refrain wafted into the dark shadows, accompanied sometimes by Mom's soft voice. The sound lifted into the cooling air as I played outside. She'd often sing "Love's Old Sweet Song (Just a Song at Twilight)":

Just a song at twilight, when the lights are low,
And the flick'ring shadows softly come and go,
Tho' the heart be weary, sad the day and long,
Still to us at twilight comes Love's old song,
Comes Love's old sweet song.

We had few possessions as extravagant as the davenport and piano; most were necessities, like our 1930s Nash. In 1952, when Harry Taylor, the junkman, drove to Minneapolis, Dad rode along to purchase a newer car. Mom, Sam, and I finished the chores and waited anxiously for Dad. He pulled up late, after dark, in a used black 1949 Ford. We considered it as good as new and *very* modern—so did Grandma Bessie. Oddly enough, although Grandpa Sam was now usually content at home, Grandma Bessie was always ready for a ride, anywhere, with anyone. After her first ride in the '49 Ford, she wrote in her diary, "Freeman's got himself a *dandy* car."

For our newest acquisition, Dad built a two-story garage into the sloping hillside, using a foundation of field stones from the piles we picked out of our crop fields every spring, then finished the second story with logs. He added attractive double-wide windows on both levels and double-doors below, painting all of them white. The lower level also had a work space and workbench; the upper level Dad used for storage. A chimney on both levels allowed for wood-heating stoves.

WASH DAY

Sometime later, Dad built another addition using lumber he cut from our trees, constructing two bedrooms and a *real, honest-to-goodness, inside-the-house* bathroom. To furnish it, Dad bought an ancient three-quarter-size bathtub with claw feet from an ad he'd found in the "used and for sale" section of the *Ladysmith News*.

My hopes had been high for a new and modern tub. Still, I had to admit it was an improvement on the galvanized metal washtub we had been using for our Saturday night baths in the kitchen, year in, year out, ever since I could remember, washing away the accumulated weekly grunge from operating a family dairy farm.

That galvanized tub and its twin served another purpose as well. On washday Monday, when I was very small, Mom became

Washday with Mom

my First Mate. But not until she finished the morning's work. She removed the washtubs that hung on the outside wall and put them in place on a homemade brown wooden bench in the kitchen. They were the rinse water tubs. Then she shaved soap from a bar of Fels Naptha into our churning Speed Queen washer, to begin the all-morning process. A chugging gas-fueled engine loudly powered the rotating agitator and belched exhaust fumes, vented through an opened casement window. Even in winter.

Mom washed the laundry in color order: first whites, then loads of colored blouses, aprons, and housedresses. All went into the same water, warmed and freshened by an occasional teakettle of steaming-hot water. After Mom sent the last work shirt and pair of denim overalls through the button-popping wringer, she tossed them in the old bushel basket she used for damp laundry. When the gaseous air had cleared through the open window, she swabbed the green patterned linoleum of the kitchen floor with a rag mop.

Even in winter, Mom hung the clothes outside. From the porch she reeled her clothesline, which was attached to a tree across the yard, and used wooden clothespins to hang the laundry. The clothes froze stiff as boards, but freeze-drying worked. Dad's full-body long-handled underwear flapped in the breeze until frozen; then Mom carried the flat and hard thin man above her head and put him and the rest of the clothes on a wooden rack behind the stove for a short drying time.

Finally came the moment I'd waited hours for: Mom flipped over the wooden bench, top-side-down on the floor, and placed it between the washtubs. With tall tub sides and high ends, the bench instantly transformed into my boat. Mom folded sailor hats out of newspaper, and clapped one on each of our heads as I saluted, "Aye, aye, Mom! We're off!" Then we sailed out onto the pea green sea, by the light of the silvery moon, with Wynken, Blynken, and Nod.

REFRIGERATORS I HAVE KNOWN

Though Grandpa Sam and Grandma Bessie had wired their house for electricity in the 1930s, they never did purchase a refrigerator—or have running water, except the cold, deep springs of Badger Creek, which also kept food cold. Just a five-minute walk from their back door, down a scenic winding path under tall shade trees, the spring had served them well for decades. When he was young, Dad caught a huge muskellunge (muskie), too large to be eaten at one sitting. After enjoying a dinner of muskie steaks, Dad sliced the rest into slabs, Grandma Bessie packed them into two-quart canning jars, and Grandpa Sam sunk them deep into the springs. All three forgot about the jars. The following summer, while dipping deeply into the stream for fresh water, the clink of glass startled Dad. The jars still held perfectly preserved fish. Classic rural refrigeration.

Freezing foodstuffs posed another problem, though obviously only in summer. Then, Grandma Bessie used Jerry and Millie's freezer. One year, when I was five, she was hurrying down the path from her house to retrieve meat in preparation for the impending arrival of Indiana relatives when she suffered an accident.

"Grandma's fallen and broken her leg," Mom told me as Dad drove away to pick up his mother and take her to the hospital.

I pictured Grandma Bessie sprawled out on the path, broken-off leg lying beside her. In my mind, I watched as Dad got Grandma into the front seat, then loaded the leg into the back seat. We'd later learn that Grandma Bessie either fell and broke her hip or broke her hip and fell. I never thought to ask Mom what really happened when someone broke a limb, so the garish image of the broken-off leg remained.

One would think refrigeration would be less treacherous for my other set of grandparents, who lived in Birchwood. But on one visit to Grandpa and Grandma Walhovd's house, I was introduced to a strange-looking machine: a white metal box on foot-high uplifted legs, tall enough for a cat to pass under, with a round metal canister, roughly the size of a small ottoman or a giant birthday cake, protruding from the top. The contraption was a compressor, part of their newly acquired gas refrigerator. On our next visit, we found it sitting crookedly in the backyard. Grandpa and Grandma Walhovd had had to be rescued from the dangerous fumes escaping from it. Since they had spent much of their lives entirely without a refrigerator, it was not a hardship for them to use their old wooden icebox until they bought a safer replacement.

Then we inherited the icebox. True, we had no ice to put in it, but its insulated walls held the cold air when we placed it in our basement, even in the hot summer. (We had no root cellar, as some did.) The icebox offered natural cold storage for most perishables. In summertime, Dad provided many meals of fresh fish, which we ate immediately. However, my parents rounded out our diet with an occasional ring of bologna or summer sausage, neither of which required refrigeration, though they did necessitate a trip to town for purchase.

Once winter's icy winds blew in, freezing was no longer a problem. In fact, Mom and Dad counted on a solid freeze to store meat in snowbanks during subzero temperatures.

Our need to manage most refrigeration by natural means, however, changed with electricity. It had first come to the neighborhood decades before, with power produced by the newly constructed Chippewa River dam near the village of Winter. The dam had created the Chippewa Flowage in central Sawyer County, with rich fishing opportunities, as well as electric power, but came with the realization that not everyone involved benefited. Despite opposition from the Lac Courte Oreilles, over the course of several years the company removed the tribe's timber and built the dam, flooding the land. The water eventually covered more than fifteen thousand acres and left hundreds of small ghostly islands.

Since the Grimh Power Company electric line ran only on Highway C, we had to pay to install it to run the half-mile to our house. Finally, Mom and Dad had saved enough money. Joe Busse, a local handyman, and his hired help spent days installing the wires and lights as we waited anxiously. They came and went on Joe's schedule, so it seemed to take forever as we listened to pounding hammers and saw wires everywhere. When they finally completed their work, Mom invited them to join us for supper in celebration. Sam and I sat at the kitchen table as Mom finished getting food on the table. Then came the big moment. Mom flipped the magic switch, bathing the kitchen in light. Instead of the dim glow of lamps, the room flooded so brightly that we were blinded and had to shade our eyes. No more oily scent of kerosene! No more glass mantles for Mom to clean and polish daily!

Soon after, Mom and Dad bought a brand-new Admiral refrigerator. On the outside, it was a gleaming white enamel box, with slightly rounded corners at the top; on the inside, it had shiny wire shelves, cool blue plastic doors, and a see-through crisper. Even butter dish and egg container spaces were carved out on the fridge door.

But the best was the freezer compartment, complete with metal ice cube trays. It chilled our cookie dough, our Kool-Aid, and a homemade raspberry "sherbet" made by beating red Jell-O with milk. Mom went to work using the local Homemakers Club's newest recipe for icebox cookies: freeze the rolled-up dough, thaw it a bit, then slice it before baking. Mom would always be able to have butterscotch nut cookie dough on hand, ready to bake for Ole when he came for a late-night visit and, as all Norwegians did, expected a cookie with his coffee.

However, the purchase of the refrigerator presented a vexing problem. Our much-used kitchen was the center of the household, with doorways leading into several parts of the newly improved house. This left little wall space to place a refrigerator without blocking pathways or protruding into the kitchen. Coming from a long line of resourceful women, Mom was not to be denied; she came up with an unconventional solution: "Freeman, just cut a hole in the wall. Put the front half of the refrigerator in the kitchen."

Alas, the backside would have to reside in my parents'

bedroom. But everyone thought the novel idea would work. Dad measured the hole perfectly. The refrigerator door opened handily into the kitchen; the unattractive motor and coils hummed noisily on the other side. Mom didn't care. Stepping back to admire her new Admiral, Mom happily announced, "Now, we'll have ice cream anytime!"

THE VOICE OF THE GOOD NEIGHBOR

For as long as I can remember, I heard voices coming from the bulky wooden box standing stock still on a shelf in our kitchen. Early each morning, I awakened to static as Mom coaxed the aged Zenith to life, spinning the plastic dials until the whirring, humming, and buzzing sound finally broke into recognizable speech: "Your farm station, eight three oh—WCCO radio, Minneapolis–St. Paul, 'Good Neighbor to the Northwest.'"

The familiar voice blasted from the upholstered speaker deep inside the radio cabinet, itself nearly as large as a small trunk. Although technically not the northwest of the United States, the WCCO motto seemed apt for our upper midwestern part of the country, and the broadcast voices certainly arrived like friendly neighbors who'd come to visit. I visualized powerful radio waves racing all the way across the flats of Minnesota to our house, bringing undemanding guests to isolated farm homes like ours throughout a wide broadcast area: Minnesota, the Dakotas, and Wisconsin. WCCO radio provided a lifeline to the outside world, offering entertainment, education, and the joy of connection to places we seldom, if ever, experienced.

When my parents finally wired our home for electricity, they also replaced the wooden radio. No more glowing vacuum tubes visible through the huge cabinet's back or powerful and large dry-cell batteries, each one as large as a giant cereal box. Like the new refrigerator, the sleek, plug-in radio of stylish ivory-colored plastic bore the brand name "Admiral." Despite its lofty name, it was hardly more in command. Reception continued to be unreliable, weather-dependent, and often riddled with static—especially during rainy spells. Just when an enterprising farmer, like Dad, most needed to hear the weather forecast to decide whether to hurry to the field for the last load of hay or to hitch up the horses—black Joe and white Queenie—to mow down another field, the radio went haywire. Then Dad had to rely on intuition and experience, reading the skies to predict tomorrow's weather.

Long before I could read, I listened to the *Noon Farm Report* and learned of "canners and cutters," "barrows and gilts," and "today's price for a bushel" of corn, beans, or wheat. I learned to listen without interruption so Dad could hear the long-range weather forecast. In fall, the weather report included early frost warnings, often predicting freezing temperatures in the low-lying cranberry bogs, warning growers when the bogs needed to be flooded to protect the ripening crop. Radio punctuated our seasons and everyday life.

While Mom baked bread, washed our dishes, or finished ironing, radio linked her to a spate of new ideas. The chirrupy voice of WCCO's own Joyce Lamont was Mom's "First Lady of home and household" and her favorite daytime companion. In addition to all of the WCCO radio fare, Mom listened to many nationally syndicated shows: *Don McNeill's Breakfast Club, Arthur Godfrey Time,* and *Art Linkletter's House Party*. Along with buddies Don, Arthur, and Art, radio soap operas filled Mom's day: *As the World Turns, Ma Perkins,* and *Days of Our Lives.*

We shared our supper with radio folk, too. The five o'clock newscast was introduced by the amazing sound of the elevator zipping thirty-two floors to the very top of the Foshay Tower in downtown Minneapolis. Though I'd never even seen a picture of the grand building or been in an elevator, I was able to imagine both. The deep voice of WCCO's legendary announcer, Cedric

Adams, followed with the news. Reports in correspondents' foreign accents reached across from distant countries I knew about only from my geography book.

On Sunday evenings, before church at eight o'clock, we huddled around the radio listening to cowboy shows such as *Have Gun—Will Travel* and *Gunsmoke* or the humor on *Amos 'n' Andy* and Jack Benny's show. Fibber McGee bickered with his partner, Molly. Detective Joe Friday caught Los Angeles criminals on *Dragnet*, and Eve Arden played my favorite character, the English teacher, on *Our Miss Brooks* at the fictitious Madison High.

On one particularly auspicious day, radio linked my tenth birthday (October 4, 1957, the day I received roller skates—*new* ones) with astonishing news of the Russian Sputnik satellite spiraling across the sky. And the day before, in New York City, the Milwaukee Braves had played, and *won*, a World Series game against the Yankees.

Whether coming from the wooden cabinet, the sleek plastic case, or the transistor radio I carried in my pocket as a teenager, radio held a place of honor for its immeasurable impact, linking us via airwaves to a tall tower more than one hundred miles away, and through that, to the world. For many years, the 654-foot WCCO tower was the tallest structure in the entire state of Minnesota, standing steadfast as a valued friend, earning the status of a trusted family member in households like mine all across—as announcers often called our listening territory—"The Great Northwest." Pilots flying over the Midwest in the 1950s would watch farmhouse lights go out after the last broadcast at ten o'clock as families turned in for the night.

THE QUEEN

A queen lived in our home. She was already there when I arrived, so I didn't know how she'd gotten there or how long she'd stood in her place. By her bulky size alone she dominated, dwarfing all else, overbearing, always present. Lest her reign should be in doubt, she proudly proclaimed her royalty in gleaming lettering across the wide girth of her white porcelain belly: a bold, blue tattoo stylishly worn like a brooch—*Monarch.*

Her cumbersome frame weighed heavily on our kitchen's plain tile floor; her once-fashionable feet ever-so-slightly and coyly lifted her heels and muscular calves. Although she displayed her crowning-glory upper torso, an ample bosom, the beauty of her once-shining surface had been tarnished by her long years of service. For, like so many of our possessions, our family castle had not been the queen's first; over time, the hands of many keepers had left their mark.

Mom had welcomed her once, as cordially as anyone would greet a future mother-in-law. She appreciated the queen's usefulness. But Mom soon learned the queen's primary drawbacks: she was old-fashioned and stubborn. By the time I was six or seven, Mom had grown quite weary of our queen's ruling presence. I could tell she frustrated Mom, taking up so much space. I sensed tension in the household. Mom was pleasant and never grumbled, but her patience wore thin. And now this decrepit kitchen companion had begun to smoke! Always at the worst time too, belching out great gray puffs of ashy film into Mom's kitchen air. I knew Mom dreamed of the day the queen would be dethroned.

But I loved Queenie for many reasons. In every season, she provided mugs of steaming milk-tea after long winter walks home from school, hot chocolate after sledding, warmth when I tucked my icy cold toes into her lap, life-sustaining loaves of warm bread and cinnamon rolls. On sweltering July afternoons, Mom and Queenie created wild raspberry jam. Queenie gave without complaint.

Still, I had to wonder about Queenie's peculiar morning rituals. I was awakened each morning to familiar sounds as she sprang into service. First, the rattle and rumble of her insides when Mom shook her down for another day, removing yesterday's ashes. Then Mom would stuff Queenie's iron belly with wood. With the kitchen warming and breakfast preparation underway, I knew it was time to get up. In no time, Queenie and Mom had eggs, bacon, and steaming pancakes waiting at the table, with fresh coffee perking.

Dad knew Mom was unhappy, but he remained dispassionate, considering Queenie's presence an economic issue. As long as she helped produce meals several times a day and radiated warmth and glow on frosty winter nights, it didn't matter that she was not in style. After all, she tolerated the wet steamy socks, heavy woolen coats, and soggy mittens he peeled off at the end of a long day. And she dried them by morning! The thought of casting her out was unthinkable. Dad decreed that she would remain.

Nevertheless, Mom's wish to banish her from the kitchen eventually came to be. I wanted to be on Mom's side, but I thought Queenie deserved a nod for her performance. For Mom, there was no nostalgia, so I remained quietly in mourning, learning one of life's lessons by taking a cue from Dad's silence: thoughts in opposition to strongly stated opinions are sometimes best left unsaid.

Thus, after faithful service for decades, we laid to rest our woodburning cooking range. Although royalty to the end, Queenie had no funeral cortège fanfare, and she did not ask for mercy. Just as well; she got none. Standing stalwartly on her heels even as she was dispatched without ceremony, her dignity crumbled; her excessive weight made it necessary to disassemble her piece by heavy piece. With the help of Mom's best ally, Sam, Queenie was exiled.

"Good riddance and off to the rubbish heap!" My mother had no remorse.

I no longer woke to early morning rumblings or felt warmth radiating from an open oven door. There was no warming oven

Sitting with the Queen

above to hold the teapot and cookie jar, no comforting sound of the simmering teakettle. I didn't like Mom's boxy new assistant: a smaller, four-burner gas range, more modern but still not brand-spanking-new. No gleam or chrome, no curvy lines or distinct design, no brooch on her belly, no ample bosom, no attractive feet and calves. Just a plain, yet fashionable, style for the 1950s.

Speaking with authority, Mom raised her spatula scepter, adjusted her imaginary crown, and proclaimed with great delight, "Now, *I'm* the Queen!"

THE HUMBLE APRON

The humble apron could be purchased or homemade and came in endless patterns and designs: plain, fancy, sturdy, frilly, or delicate. Aprons dried a tear, captured an errant baby chick, became a dishcloth or a towel, and wiped a child's dirty face. Aprons kept housedresses a little cleaner and made them last a little longer. Aprons became flags, waved out the screen door to bring the kids home for supper or to summon a hungry husband or work crew from the fields for a hot dinner at noon. A clean apron often hung nearby, to be lashed quickly into place if a haughty neighbor came to visit. The much-used apron could be just as quickly untied and thrown off when the local preacher dropped by unexpectedly during a busy day of work in the kitchen.

Mom used her apron to gather and bundle hickory nuts from the pasture as well as hold onions and radishes pulled from the garden, tiny new potatoes dug from the ground, and a lap full of fresh peapods to shell. Apples from wild apple trees, rhubarb from the patch at the abandoned Mulkey homestead, or the first wildflowers of spring—Mom carted all home in an apron.

Like many other women, Mom traded plants by tucking away bulbs or a tiny piece of a root in a guest's apron for transport home. Thriving groves of lilac bushes, orange day lilies now lining the ditches, clumps of pale purple iris, and the tiny, blue-eyed forget-me-nots all most likely began their transplant wrapped tightly in the clutch of a faded calico apron.

On slow, rainy days, Mom finally had a chance to turn to a different task. She raced to her sewing machine, reached into her stash of remnants and leftover fabrics, and whipped up potholders—or expertly remade an old work shirt into yet another apron. For decor, she chose lace, rickrack, bias tape, or something from the button box. Perhaps she added a decorative pocket, cleverly adding a touch of flair. With creativity, women like Mom made patterns and designs as unique as the housewives themselves.

Mom favored a utilitarian apron that covered her upper body, circled her neck, and tied around her waist. Flimsy, organdy aprons covering only the lower body provided no real protection at all and couldn't be used for mopping up spills. However, even farm women wore them when serving at weddings, pouring coffee at showers, and at Christmas or holidays, decorated with seasonal themes.

Mom, like many other women, kept an apron or two hidden away for a quick, small gift, such as for an unanticipated birthday. Checkered cross-stitch aprons made fine gifts; cobbler aprons with many pockets were useful for women and kids alike. Women welcomed the gift of a well-designed, built-to-last apron on any occasion. One more apron was never too many.

THE KITCHEN TABLE

Our kitchen table wasn't much. Its hefty oak legs creaked from constant use. When the table began to sway, they had to be screwed more tightly into the solid oak top. Mom brightened the table annually by purchasing a cheery, patterned oilcloth each summer, but it soon faded. The hand-me-down table from the Walhovd side of the family had spent years as the center of household activity: work, study, entertainment, conversation, and of course, eating.

Across the kitchen table, Mom passed pancakes, bacon and eggs, oatmeal, and coffee; bowls of potatoes and gravy, string beans, boiled beans, soup, sauce, and stewed tomatoes; dishes of pears and peaches; loaves of homemade bread, buns, cinnamon rolls, and slices of pie; platters of farm beef, roast and fried, direct from hoof to table; frothy farm milk, straight from cow to pitcher. Three square meals every day.

On the table, Mom cooled jars of canned tomatoes, raspberry and plum jam, and quarts of freshly cut corn from the cob. Mom and I, and sometimes Sam, sat there to shell mountains of peas, snap beans, peel potatoes, and clean and sort wild raspberries.

At the table, Mom gave haircuts, darned our socks, patched overalls, snipped stiff pieces of crisp cotton for aprons, and laid out cloth and cut patterns for sewing housedresses, shirts, and blouses.

It was at the table that Mom finally sat down to rest each

afternoon with a teapot and a cup of hot Lipton tea. She poured one with milk in it for me; or, in summer, she poured cherry-red or lime-green Kool-Aid. Mom then sometimes slipped away to produce a hidden bag of lemon drops or a Hershey bar to share.

The table held the parts of Dad's shotgun, deer rifle, and .22— taken apart, oiled, cleaned, and polished, to get his guns ready to hunt for partridge, ducks, deer, and smaller animals. When rain poured in sheets, weather so drenching it was unsuitable even for good fishing, Dad emptied his tackle box onto the table to sort fishing lures and "spoons" and rearrange multihooked muskie baits, wooden plugs, imitation rubber angle worms, and jars of pork-rind bait, shaped and dotted like frog legs.

On the table, Sam and I pounded modeling clay, smeared paper with homemade finger paint, and gripped drawing pencils and crayons. We fought over which of us would get to draw pictures on the brown-paper catalog covers. We used them to save money on paper purchase and because they were large and heavy. Sam fussed over gluing pieces of model airplanes just so and carefully daubed paint-by-number kits. When kids came to visit, the table held the net for a wild game of ping-pong. We played crazy eights, old maid, Cootie, dominoes, Mr. Potato Head, and my favorite game, Uncle Wiggily.

Sitting at the table, Mom shared riddles and rhymes and read aloud from our storybooks, collected poems from Robert Louis Stevenson, and our kids' Bible storybook. Sam and I tore out the 1950 color-splashed pages of Dr. Seuss's first printed stories from *Redbook* magazines, begging Mom to read "The Big Brag" and "Tadd and Todd" again and again. She did, until they were worn to shreds.

At the kitchen table, Mom taught us songs, repeating the verses line by line until we got it. One evening she sang out, "There's a hole in the bottom of my drum, there's a hole in the bottom of my drum, there's a hole in the bottom . . . there's a hole in the bottom

. . . there's a hole in the bottom of my drum." Easy enough! I was sure I had this one down cold, loudly singing, "There's a hole in the bottom of my rump, there's a hole in the bottom of my rump, there's a hole . . ." Before I hit the third "rump" Mom and Sam had already collapsed into gales of laughter, making me furious.

Mom and Sam pored over his high school homework: *The Merchant of Venice*, geometry, essays. I studied spelling, arithmetic, and, at the insistence of my fifth-grade teacher, Mrs. Vitcenda, recopied tablet sheets of my handwriting until I was cured of sloppy writing. In high school, with my portable typewriter, I practiced typing and banged out book reports and term papers.

Mom and Dad spent hours at the kitchen table doing sums, running the household, and writing. Mom recorded farm expenses and compiled orders; besides clothes and shoes, she sent off for seeds and film development. Every Sunday after evening church, Dad sat down to count dollar bills and record the figures from the collection plate. Dad and Mom's letters to relatives took shape on the table.

When company came calling, which was often, they never left without a meal. Ole joined us at the table in the late evenings for some of the best times. Our family also became closely entwined with the Miltimores, Rob and Ina—or as we called her, "Inee." Mom and Inee became fast friends while Dad and Rob spun stories over the table. Despite a twenty-year age difference, Rob and Dad were drawn together by their interest in the outdoors. They often fished together from a canoe, although neither knew how to swim, and remained hunting partners for many years. Dad knew that few men alive had Rob's experience, that Rob's tales were not exaggerations, and he valued the older man's wealth of knowledge.

Indeed, Rob and Inee rumbled up and over the hills to our house in their snout-nosed '40s Ford pickup with such frequency that Dad gave it a name. When we heard the familiar hum of their brush-painted truck coming, one of us yelled, "Here comes the Green Hornet!" We were always happy to see them bounce down our lane. They often joined us for a meal, but on occasion, Rob would reply to the invitation, "Thanks, but we've already et."

For lots of company, Dad and Sam pulled the table from the wall to provide more room, and I helped Mom gather every chair, even the piano stool, and Mom's laundry bench. Kids, perched on the bench behind the table, were lifted up by sitting on stacks of thick catalogs.

Under the daily burden of food, figuring, and frivolity, our table collapsed more than once, weary of the load any piece of furniture would struggle to bear. Its legs gave way, folding under when we neglected to shore it up. The table became swayback with so many rounds of company requiring extra leaves for loads of food and games after the meal—until Dad added metal bars underneath for reinforcement. In other words, support.

Support. Smiles. Joy.

News—happy and heavy.

Challenges. Disappointment. Deaths.

Change.

A CAT'S WORLD

In one photograph, She, the cat, wears a serene look of peaceful slumber. The mellow, tri-colored calico seems to purr blissfully. She was tucked into my narrow doll bed, a bonnet crookedly clamped down on her folded-over ear, but She didn't seem to mind. She'd been mauled and hauled around since I was a toddler. Other photos show She stuffed into a pink jacket, wedged in a doll buggy, and robed in clothes that were meant for a doll I'd discarded long ago—in favor of furry feline flesh and blood.

Cats came and went on a farm. Stray cats wandering farm country faced a tough life. Though it was not their preferred elegant lifestyle, cats settled for a home in the barn, if only the farmer allowed. Some were crafty enough to earn their way up to indoor housecat, which, for a cat, was a worthy accomplishment. Dad knew cat behavior. "Ya know, Peg," Dad warned, "cats're patient creatures, they'll take their time makin' their way into yer house, one paw at a time."

Such was the case with my first calico cat, whose real name is lost, and Tommy, a splendid yellow tom. Tommy and Calico both "ran away"—a sad euphemism for any number of natural disasters that may happen to a cat on a farm, none of which you'd want to know more about.

Dad, like all farmers, was faced with the dilemma of a skyrocketing cat population unless he took measures. Dad loved cats but had to be pragmatic in culling and humanely limiting their number, especially the females. We fed and provided for our working barn cats: mousers that controlled the rodent population. They patrolled the barnyard, occasionally leaving the harvest of their hunt on our doorstep to prove worthy of continued room and board. As the most avid cat-lover in the family, I was given the task of naming our mousers, so I cleverly color-coded them: Blackie, Whitey, and Grady.

Tabby, an alert, striped tiger kitten, rode home in my lap one day when I was ten and my mother had gone to the hospital. Our insightful neighbor had made sure I had a distraction. Tabby became my constant companion. He slept on my bed each night, unless lurking and stalking, and set his cat clock each afternoon to greet me on time at the end of our dead-end road after school. He grew into a handsome, strapping tomcat that regularly ventured out, catting around. An unaltered male, he spent many days and nights prowling, arriving home scratched and battered after defending his honor and his farm country territory from cats who didn't belong—as well as from assorted other creatures of the night. Tabby weathered a vicious owl attack that left him with

Peggy and Tabby

a mangled jaw and chin. As a result, we fed him soft food for the rest of his life.

Early on, I wondered if Tabby could swim and foolishly threw him into the pond to find out. He could. He emerged bedraggled and sodden, his cat ego in tatters. He was furious and ran away deep into the woods, not even looking in my direction to give me the searing look of disgust I deserved. I was immediately worried and spent sleepless nights, heartsick when Tabby did not return. But he forgave me when he nonchalantly strolled in a couple of days later and crawled onto my bed to sleep as usual.

Years later, I rescued a tiny kitten, another cat I named Tabby, who soon came to be Dad's cat. The pair were perfect companions. Dad taught his Tabby to salute for a favorite treat. On command, Tabby sat on his hind legs like a dog, lifted his right paw to his eye and twitched it—in a salute as perfect as a cat ever made—earning a sip of the evaporated milk Dad added to his coffee.

Tabby—and Tabby—taught me that cats show loyalty when it is deserved (and sometimes even when it isn't) and are models of composure, living by Nan Porter's quotation: "If cats could talk, they wouldn't."

It wasn't only the domesticated kind that walked around our farm, though. One evening, Mom became aware of a strange sound above the suddenly-changed crow calls and the twilight melody of robin's song. Coming across the freshly mown, forty-acre hayfield was a repeating throaty cough. She strained her eyes toward the hoarse, unsettling sound, now louder than ever, moved to a stack of lumber Dad had piled nearby, and climbed to the top for a better view. From her perch she watched a tawny cat, much bigger than a bobcat, with a long tail trailing behind. The animal slinked warily across the field, as if in stealthy pursuit of some unseen prey. Though it paid no attention to Mom, she didn't need to see more. She called for Dad, but quickly realized it was unlikely he would hear her above the din of the milking machine. Mom's dash toward the barn was halted by a heart-stopping scream. Mom called it the scream of a murder victim—a sound she would never forget.

Dad had never seen a big cat, but he *had* seen the large footprints and heard reports of cats from Rob and other old-timers. He did not doubt Mom's sighting. The account of the cat brought a buzz of excitement across the neighborhood. Some nodded wordlessly, as if the tale struck a chord, while a look of skepticism crossed other faces. But all were entertained. "It sure as heck was the Wampus cat," Mom joked.

Gone Farming

PICKING AND PRESERVING

Mom's home canning of our produce began in early spring with rhubarb, and it continued throughout the next four or five months. However, we didn't can everything. We consumed as many vegetables as we could, direct from garden to table. The garden was organically fertilized with composted cow manure—though no one would have described it that way back then. Garden harvest was well underway in June with radishes, lettuce, and green onions. One of the rites of summer was the first meal of tender young peas, tiny carrots, and the first new potatoes, cooked together in a white cream sauce.

Shelling peas, perhaps the most tiresome of all tasks, came first. It was unlikely anyone would have taken so much trouble had the work season not just started. Next, we picked and cleaned row upon row of snap beans. Only then did I realize that through the rest of summer something would always need to be picked, cleaned, and, usually, canned. Later in the year, our veggies formed a kaleidoscope of color: quarts of deep red beets, jars of orange carrots, yellow beans, green beans, and pale green peas.

Then came my favorite: pickled cucumbers of every variety filled our basement shelves: sweet dills, sour dills, garlic dills, bread and butter pickles; and bright green, cherry chunks, made of larger cukes. I loved pickles of any kind, which was good because Mom didn't stop at pickling cucumbers. She made green bean pickles, crab apple pickles, spicy beet pickles, and even watermelon rind pickles. Pickle relish, too. It seemed Mom's pickling possibilities were as endless as the cucumbers. The season came to an end when we left monstrous, yellow cucumbers, nearly as big as watermelons, to rot away in the garden.

In the summer, my harvest began with the first hint of sweetness when, returning from my exploration of the cow pasture, I found a handful of tiny, wild strawberries. By the Fourth of July, the first wild raspberries ripened. Lured by Mom's (meager) pay of ten-cents-a-quart, still I persisted. I stumbled over rock piles infested with hornet nests, clouds of swarming mosquitoes, and sunning snakes. Confronted with prickly briar bushes, gouging thorns, stinging nettles, spiny, scratchy thistles, and stink bugs, nothing stopped me from plunking the juicy berries into my bucket, even when they smashed into a blob, or I tripped, spilling the bucket while running from a disturbed bumble bee. The payoff? Mom's raspberry pies and canned raspberries, a winter treat we all enjoyed.

When there were no more ripening red berries, I took a welcome break, resting and healing my superficial wounds until the equally prized blackberry season began in August. To reach those berry patches, I knew I would have to contend again with the same obstacles, but I also knew that I could soothe most itches, especially from stinging nettles, with the crushed juicy leaves of orange jewelweed. Jewelweed also provided entertainment: fat and puffy seed pods would explode at even the slightest touch, hurling the seeds into soil to live and grow again, or into hair, pockets, or berry buckets—no wonder we called this plant by its colloquial name, "touch-me-not." Yes, summertime berry-collecting had its distinct advantages, not the least of which could be seen on our basement shelves.

Next to the jars of produce and pickles, Mom stored all her jellies and jams, topped with paraffin. Purple plum and apple jelly, marmalades, blackberry and raspberry seedy jams, all just waiting for use on winter toast. Then came rows of whole fruit to stock our winter cache. Mom's canning season was not complete until we'd brought home wooden crates of juicy fruit from Veness's: freestone peaches, yellow pears, and whole plums. Mom even made a kind of sweet tomato jam, called "preserves," which we also spread on toast.

Whether categorized as a fruit or vegetable, tomatoes took up a surprising amount of space on our shelves. In the northern climate, tomatoes ripened slowly. But once the ripening began, they arrived by the bushel for a *long* time. When Mom canned tomatoes using the wood-fired cook stove, it didn't take long for the kitchen to become hot and steamy, all the more so in the heat of August.

Mom ladled the whole peeled tomatoes into jars, lowering them carefully into the blue-speckled canner for a boiling-hot water bath. She also used a large pressure cooker; the noisy rattle of its lid echoed as it hissed steam, and the kitchen took on the sour, acidic odor of tomato skins. I dropped a hot lid on top of each jar and loosely screwed on the top. Then we listened for the "ping," assuring us that the lid had sealed. After the jars cooled overnight, my job was to safely carry the tomatoes to the rapidly filling basement storehouse.

One winter day in 1956, with the coming of the first early snowstorm, I knew for certain we'd moved on to the long winter season. When school was out, Sam and I tramped home through the dense swirling flakes, surprised upon arrival to find our hay wagon sitting alongside the basement windows. Mom and Dad had hurriedly picked the remainder of garden bounty, loaded it atop the wagon, and brought it to the house. They'd opened the wide windows in the stone foundation of our house for quick access. Peering inside, we could see a rainbow of mountainous produce: knotty, green Hubbard squash; smaller varieties of golden butternut and buttercup; and orange pie pumpkins. Leafy cabbages extracted from the ground, clay still clinging to their roots; cream-colored rutabagas; and long green sticks of Brussels sprouts. Bushels of earthy-brown russet potatoes were stashed in the bin beneath the narrow stairway; yellow onions hung in reused net bags. I knew orange carrots had already been buried in crocks of dry sand, a way to keep them fresh. There were buckets of red apples: Duchess, huge Wolf Rivers, and what my grandparents called "Prairie Spy," for winter cooking. The parsnips, however, still snuggled under a bed of snow in the garden, waiting to be dug from newly thawed ground next spring when they were at their best, fresh and sweet.

FISHING AND HUNTING

Spring, which would seem so far off yet, was still able to fill me with anticipation. Ah, spring, when the suckers "ran," caught only in that season's cold water while spawning each April in local streams. Rob would line his homemade outdoor smoker with them, creating flavorful slabs of smoked fish. Using her own recipe, Inee would drench the suckers in a salty brine that softened the bones and preserved them. We thought the flavor surpassed the canned salmon we bought at Veness's.

At that time of the year, Dad, Rob, and Ole also took seine nets to Lake Superior, returning with gallons of tiny fish for smelt fries. Brook, rainbow, and brown trout from Badger Creek and nearby lakes and streams also fed us in the summer. Dad's fishing trips to Deer Lake and the Chippewa River and Flowage supplemented

Dad and Ole during hunting season

our diet with plenty of impressive muskie, northern pike, walleye, and bass.

In the fall, Dad also provided another source of protein for our table—venison. He made it a point to be ready for winter before the beginning of deer hunting season in mid-November. Dad often proclaimed, "Huntin' season usually begins in the warmth of fall, but ends ten days later, in the depths of winter." This year, "usually" would prove correct. Indeed, this early snowstorm did not end until a foot covered the frozen ground and the temperature dropped well below zero. And, in a way, Dad welcomed it. "The snow'll help us track a wounded deer come huntin' season," he commented. Most likely we'd be eating venison pretty soon.

Dad had been waiting for the first big snowfall for another reason, too. The winter task of butchering beef was delayed until we had a good cold streak, although a fattened steer was always ready in October. The animal would meet its bleak fate and become our winter meat supply along with the venison. Dad and Mom packed the rock-hard packages of frozen roast, steaks, chuck, and soup bones into ten-gallon milk cans, and when snow arrived, they buried them deep in a snow bank. After winter's end, we took any remaining meat to the nearby town of Ojibwa, where we rented freezer space. Sam and I looked forward to these trips when we could visit the open-air shelter at the Ojibwa Club. It displayed an original Native American birch bark canoe and a dugout canoe and also featured a sorry black bear pacing his cage beside the Chippewa River, perhaps wishing he could be joyfully released just as spring had released us from winter.

But on this first snowy taste of the season, we could not hope that winter would truly be over for another six months. Dad had already filled the woodshed with split logs of oak, maple, and hickory from our woods, ready to make a cozy fire. In case we were snowed in, Mom had added a few extra basic supplies to our cupboards: sugar, salt, cocoa, coffee, and tea. With fifty-pound sacks of Robin Hood Flour and blue-striped packages of compressed yeast, Mom's trusty oven would produce brown loaves of the bread she baked several times a week. It was heavenly in any kind of weather, but especially in winter, slathered with lots of butter and wild, seedy raspberry jam. Supplied with milk and cream from our dairy cows and cheese and butter from the Ladysmith Co-op Creamery, we would take on the approaching winter.

As Sam and I peered through the basement windows after school, we saw the fruits of Mom's labor stored on the shelves that Dad had built of timber harvested on our land. We knew that as winter made its way down from the Arctic, frigid winds blowing in, mercury plummeting, snow piling high, and long, silvery icicles hanging from our rooftop, we would be safe, fed, and warm. Sheltered from the north wind in our secluded valley, we would hunker into our log house, atop our storehouse of riches.

Inee's Preserved Sucker Fish

1 teaspoon salt (per quart jar)
2 tablespoons vinegar
2 tablespoons olive oil
4 tablespoons catsup

Soak fish overnight or several hours in cold salt water. Pack sliced fish (cut in chunks) in quart jars and boil for 6 hours in hot water bath, or 3 hours in a pressure cooker. Let sit for a few weeks before opening sealed jar. Bones of fish will soften.

COW CULTURE

When Sam was small, Mom and Dad grew rutabagas and red mangel beets for cow feed. Both highly nutritious root crops grew well in the cool fall temperatures and stored well, becoming sweeter as the season wore on. Dad cut the ten-pound roots into cow-bite-sized pieces. Sam remembered our cows chomping happily on 'bagas and beets, as pink frothy juice spilled through their grinding teeth and dripped into the manger. We all loved 'bagas, too, especially, as Sam and I called them, *smashed* 'bagas.

There wasn't much that Sam and I didn't know about our dairy cows. They bookended our daily existence and kept the pace of our lives. We learned all about cows, calves, and heifers—two-year-olds ready to breed—by observation and by playing and working around them. Most of our herd were brown and white Guernseys, known for producing milk rich with butterfat—literally a farmer's bread and butter. We also had a handful of black and white Holsteins, bigger cows who gave more milk that was far less rich in fat. My favorite cows, maybe because they looked like teddy bears, were our two Brown Swiss—Brownie, who led the herd, and Swiss Sister, who still had her horns. For safety, Dad removed most of our calves' horn nubs. However, docile Swiss Sister was not a threat, even with her horns.

Cows carried the personalities of their breed: Guernseys were gentle, like Brown Swiss; big-boned Holsteins pushed their weight around. But individual cows had personalities, too. Like people, cows could be calm and complacent or ornery and stubborn; some cows bossed other bovines around; some were bright natural leaders, others timid followers. They could be social, independent loners, or outcasts. Like chickens, cows had a pecking order. Cows stayed away from bullies and recognized well-behaved leaders for their intelligence, experience, and confidence in keeping the herd in check.

It seemed as though cows had their own conversations about humans; sometimes they appeared afraid when they did not understand what we wanted them to do, or, maybe especially, when they did: "You want me to go across that muddy culvert? With that strange, hollow sound underfoot? *I'm* not going until Brownie goes."

Cowherd leaders came in handy in the barn, as well. When Dad brought the first-time-mom heifers into the barn for milking, they had to be taught the rules of cow etiquette. Creatures of habit, cows routinely returned to the same stall. The look in an offended cow's eyes revealed her distress when a heifer entered the wrong stall. The upstarts caused chaos until they eventually found their places in the barn and herd.

And don't think humans can get away with mistakes or with treating a cow badly, either. Cows recognize faces even long after an event. They also remember their babies and the locations of shelter, water, and the best places for grazing.

Although our cows often grazed in the pasture, some of which was wooded, Dad never let them in the trees when acorns first dropped from the oaks. A diet of green acorns could limit or dry up milk production, as could a number of other woodland plants harmful to livestock, such as common milkweed, horse nettle, and black nightshade.

Likewise, Dad was careful not to let the cows out to graze in the hayfields when morning dew still covered the mix of pink and white alsike clover, larger red clover, and Timothy grass. The sweet clover self-seeded and created a perpetual gathering place for bees hovering in the lovely fragrance of the hayfields. Despite its bucolic appearance, however, clover could be deadly. Ingested wet clover could cause a cow to bloat, resulting in a downed animal if the pressure on her stomach was not released.

Our herd's health meant our livelihood, and this was reflected in our kitchen medicine cupboard. Some prescriptions seemed to work perfectly well for both cows and people. Sulfite powder, for instance, sprinkled on wounds a cow might get from scraping on barbed wire, worked fine for Sam's crashed bicycle road rash or on my knee scrapes after a tumble chasing my baseball. Likewise, the miraculous Udder Balm for Bossy's sunburned bag—when she slept in the sun with her bag exposed—worked just as well for our berry-picking sunburns.

The herd kept us busy and provided our main income, entertainment, and food. Dad bred some cows with beef animals, producing coal-colored Black Angus or red Hereford calves with cute white faces. Sam and I loved to play with the good-natured little animals. And when they were old enough to go to the pasture in early spring, like the rest of the herd at that time of year, the calves frolicked and jumped like children, delighting us. They didn't know that in two years they'd be full grown beef and ready for Dad to butcher.

Our cows seemed to have an innocent look in their eyes, as though they just wanted to quietly exist in their own peaceful world. Like me. Lounging in the summer sunshine, chewing their cud and soaking up warming rays, our cows truly looked content.

MILKING

In order to produce milk, cows had to birth a calf once a year, in late winter or early spring. Annually, all of our cows usually successfully birthed one calf—we never had twins, but our neighbors did. We knew a cow was ready to breed when she began to act strangely: long bellowing moos, restlessness, ear flicking, a wild look in the eyes, and the tell-tale mounting onto the backs of other cows.

Once, on a ride through the country with our city cousins, one of them spotted mounting cows in the pasture and piped up with, "Look, Mom! Circus cows!" Sam and I rolled our eyes, knowing better, but not informing them.

Gestation in cows is 283 days, just a few days more than for humans. When a cow lumbered around clumsily, we knew she was due; Dad knew the importance of being aware of this sign in order to keep the mom-to-be enclosed in the barn. For if a cow gave birth in the pasture, she would disappear with her newborn and could become belligerent, even dangerous, to the farmer she knew well—especially when he attempted to retrieve the calf from the protective mother's hiding place.

The mother's colostrum, the first three or four days of milk, could not be sold to the creamery because it contained bacteria, but it was nutritious feed for a new calf. We fed the calves first with this, then later with a powdered mix. The Guernsey calves, with soft, curly-haired brown and white coats, would eagerly reach to lick my hand with a sand-paper rough and sloppy tongue, after I'd immersed it in warm cow's milk. This would entice them to learn to drink by putting their head in a bucket, rather than our having to feed them with the time-consuming "nipple pail"—a bucket with a rubber nipple attached. Since calves butt their mother's teats (we called them "tits") in order to bring down the milk and get it flowing, sometimes calves would suddenly butt the bucket and spill milk all over if we weren't careful.

While calves learned to get nutrition without their mothers, a heifer was first learning that humans would take her milk as well as her calf. The first milking machine attempt frightened the animal and put the farmer in a dangerous position. Every farmer carried scars or broken bones from when a heifer inevitably tried to kick off the machine. Usually she became accustomed to the udder cups dangling from the stainless-steel contraption, but sometimes the young cow didn't learn to comply. Most farmers then wasted no time in "shipping" her (shorthand for *hamburger*).

Our lives were entwined with the rhythm of the herd, season by season, with milking on a daily basis at five o'clock in the morning and at six o'clock in the evening. Dad was present for both, and sometimes our family of four required all hands on deck, especially in the evening, for the one to two hours milking and related chores. I remember just one day when Rob milked the cows, allowing Mom and Dad to take a jaunt to the Minnesota State Fair. Even then it was not an overnight trip; they left early, after the morning milking, and returned late the same night. In addition, daylight saving time threw both the cows and farmers out of whack. Dad grumbled that he had to get up and search for the cows in our pasture in predawn darkness. The cows had to reset their internal clocks, as well.

For many months of the year, milking time began with bringing the cows back from whichever pasture they were grazing in. We moved them after a week or two in the same area once they'd eaten the grass down low. Often it was a challenge to move cows from one pasture to another; we called, "Come Boss, come Boss," to entice Brownie, knowing the rest of the herd would follow. A bucket of tempting grain didn't hurt, either.

While some farm kids had milking and chores before school

and arrived in smelly barn clothes, in our family, Dad took care of the morning milking on school days. In summer, often Sam or I fetched the cows in the late afternoon and corralled them in the barnyard to await milking time. Then, while Sam and I herded the cows into place and closed the stanchions, Dad readied the milk cans by placing atop each a giant stainless-steel strainer, shaped like a funnel. Then it was time to get down to milking. Usually. Sometimes we had to hold our cows in the barnyard while getting another load of hay in before rain came. We'd hear them mooing in complaint; they wanted to be relieved of their bulging bags.

To keep the milk safe, one of us mixed a strong disinfectant, called BK Powder, with water, then carefully washed each cow's bag and udder. Mom, Sam, and I took turns sitting on the milk stool—the handiest tool in the barn. Dad had made our version: a lightweight, three-legged model, legs driven slantwise into a wooden slab. Over the years, many backsides sliding over it had worn it smooth. We lifted it easily in one hand while carrying the disinfectant wash pail in the other and swung the stool into place, on the side of each cow's business end. A careless washing job showed up in the monthly report from the creamery, which tested for bacteria. A farmer never wanted a string of bad reports. Sam and I dreaded getting the cows from pasture in the summer only to find they'd cooled off in a mucky area of the swamp, their hooves, heels, bags, and udders coated in layers of hardened mud. It seemed like it took forever to painstakingly wash it off.

Mud wasn't the only thing accompanying cows in the summer. An unending battle with flies went on in every barn; flies often rode cows' backs into the barn at milking time. Luckily, a cow carried her own flexible fly swatter on her backside. A cow bothered by flies crawling across her hide first twitched her skin. If that didn't dislodge the pest, she sent her long tail, a curly mop of hair at the end, flicking toward the disturbance. Every one of us seated beside a cow had been swatted in the face when failing to dodge a flying tail—wet, smelly, and none too clean, but not as bad as another farm hazard. When a cow suddenly arched her back and lifted her tail, we wisely moved away—fast. Cows could flood the gutter with a stream of urine, or worse. Fresh green grass and plenty of water in summer brought loose green manure splatting. We learned the hard way never to stand behind a cow without watchful attentiveness.

After we scrubbed each cow clean, Dad hung the milking machine on a strap around the cow's waist and placed udder cups on each teat. The engine pulsed to draw milk from the bag. We listened carefully to the sounds of our two Surge milking machines; when a cow's bag neared empty, the sound changed. If the milker ran too long, it could damage a cow's udders. Then Dad replaced the machine, taking his turn sitting on the stool to milk by hand—or strip—the cow of her last bit of milk, streaming it into a pail.

Cow after cow, Dad and Sam carried the full Surge machines, now made even heavier, and poured the two or three gallons of milk into the wide-mouthed strainer. Milk flowed slowly through the clean white filter and dripped into the can. Dad regularly inspected the pad for particulates—bits of hay, mud, or even a stray fly—to make sure we were doing a good job of cleaning and disinfecting the udders. In summer, four or so ten-gallon milk cans were filled to the brim, fewer in winter.

In summer, after the evening milking, Dad took the still-warm cans of milk to our cooling tank, which was filled with frigid water from deep in our well. It kept them chilled overnight. In winter, an opposing strategy became necessary. On frigid nights, Dad left the cans cooling outside until bedtime, then brought them into the kitchen. If left out until morning when Frank, the milkman, arrived, the milk would be frozen solid.

While Mom and I fed the calves, housed separately from their mothers in another shed, Dad and Sam busied themselves with cleaning out the gutter in the barn with a shovel and five-pitched manure fork. The manure would not go to waste. In spring, Dad would pitchfork this natural fertilizer onto what would be our crop fields of oats, corn, and hay, planted in rotation. He also added lime to our soil and, now and then, commercial fertilizer.

In winter, the chores increased, as Dad or Sam had to pitchfork loose hay from huge haystacks piled during the summer and add oats, before placing it in the manger. Morning and evening, each cow also received a large scoop of corn fodder, chopped by Dad's hammer mill. Each week, he loaded shocks of corn from the wintry field onto a wagon, and our team of horses pulled it across the frozen corn stubble of a snow-covered field. As winter wore on and our supplies decreased, Dad had to purchase feed. The fifty-pound bags were a mix of grains with a dollop of molasses in them—cows have a sweet tooth, too. Molasses was cow candy and provided nutrition, as well.

Finally, in every season, Sam and Dad swept the barn clean and, in winter, pitchforked any uneaten remains of hay into the stalls for the cows' bedding. In summer, after milking time, we simply released the cows to one of our pastures to graze, then Dad and Sam tossed lime on the floor and sprayed the barn with DDT to help rid it of flies.

Whether post-morning or post-evening milking, Mom, Sam, or I performed the remaining task in our kitchen: washing all of the parts of the milking machines. We used a brush to thoroughly scrub everything with a strong cleaning solution.

After the evening milking, *at last*, between eight and ten o'clock, we all could relax before turning into bed, to be ready for another milking shortly after dawn the next morning—when it would be time yet again to rouse the cows from slumber, whether in the barn or sleeping soundly in our peaceful northern Wisconsin pasture.

A HORSE'S LIFE

Joe, our long-faced workhorse, had frisked across our farm for as long as I could remember. Though still spirited, his best days were over by the time I was twelve. Joe had had many changing team members in the pasture with him over the years—dapples, sorrels, roans, and white horses—friends, I suppose, who had left while Joe remained. Some of them, I knew, had died of old age and been carted off to a fox farm where their carcasses fed animals raised for their furs.

Sometimes I went to the pasture with Dad when he went to get the horses. He would rattle the handle of a battered metal bucket with a treat of grain, enticing Joe forward. Munching grass, Joe would lift his lanky head when he heard the bucket and Dad's voice gently calling, "Come Joe, come Joe." Joe would sprint closer, as if to say, "I'm coming, I'm coming." Dipping his long nose into the bucket, Joe would happily chomp the grain with his pearly teeth. Dad would slip the bridle over Joe's ears and the bit into his mouth, leading Joe and his white partner, Queenie, to the barn to harness them to equipment for an afternoon of work. Though Dad used the 1940s Case tractor for most farm jobs, the

Old Joe in the pasture

horses still cultivated corn, mowed hay, towed a sloshing tank of maple sap through slushy snow, skidded logs in wintertime, and dragged a heavy load of rocks from our rock-strewn fields on the stone boat. With muscles rippling and sweat dripping, the team often strained forward.

It would be many years before I knew Joe's whole story. Jerry had bought Joe at what seemed to be an especially good price for such a large, strong, young animal. However, Joe proved to have anything but a mild temperament, and Jerry's attempts to tame him had been futile. Joe remained feisty, aggressive, and unpredictable—a half-ton bundle of raging equine energy and anger. Wild Joe looked like a gelding but behaved like a stallion.

Dad had checked Joe's teeth and hooves, and upon close examination of the horse's "plumbing," Dad realized that Joe had a reason to be distressed; he was a frustrated young colt with retained testicles, or in vet's language, cryptorchidism. In other words, Joe was unable to release testosterone. After checking with a local horseman, Dad felt certain he could remedy the problem, so he made Jerry an offer he couldn't refuse: "I'll pay ya the money ya paid for Joe. An' wager fifty dollars I ken get 'im tamed and in harness within a month. Then he's mine."

Jerry scoffed, "#*!%*!"—or words to that effect—sure that Dad had lost his mind. Jerry's snorts of laughter echoed in Dad's ears.

Thankfully, I do not know the details. Dad succeeded in turning Joe into an energetic and vigorous colt. A few weeks later, Dad triumphantly led a harnessed and docile Joe for Jerry's inspection. After collecting his wager, Dad disclosed his discovery. Jerry was not pleased at being upstaged by his younger brother, although each of them used Joe over the years.

And Joe? He seemed forever grateful, even loyal, to Dad; that horse would ramble about in his pasture for many years to come, well into retirement.

Family, Friends, and Neighbors

VISITORS

Though it seemed like farm life isolated us, a surprising number of people came through our door, and they often arrived unexpectedly. While we worked in the garden, house, or barn on the hill, our ears stayed tuned, listening for the sound of tires crunching or the sound of an approaching engine in rattling cars, beat-up pickups, or delivery trucks. For whatever reason they came—to perform services, deliver, take away, enforce, sell, or simply visit—most brightened our days with their appearance.

Some provided expertise, such as the Surge repairman and the veterinarian who arrived to care for an injured or sick animal, or in some cases, to prevent injury. Although Dad cautioned us never to leave bits of metal around, as a cow might accidentally ingest sharp pieces left in the pasture or hayfield and perforate one of her four stomachs, Dad wouldn't take any chances with this potentially serious hazard. The problem was solved with a three- to four-inch-long "cow magnet." The vet used a "balling gun" to shoot the powerful magnet into a cow's mouth, forcing her to swallow it. (After butchering, farm kids traded stories on the amazing array of metal objects still attached to magnets—bits of wire, nails, staples, paper clips, bobby pins. Then the magnet often found its way into kids' pockets where it continued to entertain.)

Other people made deliveries. Frenchy and his wife, Helen— the only woman who arrived for business—supplied tractor gas from a red Texaco truck. In late winter, the feed store delivered grain for the animals, and in spring, lime and sometimes fertilizer for the crop fields. Most unusual, the artificial breeder serviceman brought the frozen semen to inseminate our cows when each went into heat, or estrus. Estrus usually occurred in one cow at a time over a period of a few months in the summer. Then local people, trained as technicians of ABS (American Breeders Service), wore long plastic gloves over their arms as they inserted semen into our ovulating cows.

Some folks, like Frank, came to take away. Harry Taylor carried away rusty equipment to sell for a few cents a pound; George Busse, the shrill-voiced calf buyer, transported the little bull calves to market. Born to their dreary fate, male calves had no value in a dairy herd. Likewise, when a cow became too old to produce much milk, she, same as the uncooperative heifers, was shipped, driven away by the cattle buyer. In the winter, logging trucks hauled Dad's timber to be sold and sliced into lumber.

One man came to enforce regulations. Sam and I most dreaded the visit by the creamery milk inspector who always, despite our diligent cleaning and rinsing, seemed to find a bit of stubborn residue remaining on our milking equipment. Then Dad would lecture us about how we had to scrub more carefully.

But most men came to sell. Some hawked wares of little use or peddled things we didn't want or couldn't afford: magazines, insurance, sets of encyclopedias, and accordions and lessons for kids. Some salesmen sold barn whitewashing and painting services. Dad hired some and declined others. Most salesmen accepted no for an answer. One did not. Dad, who became hard of hearing, had once made an inquiry about hearing aids. He was definitely *not* interested but continued to be revisited by a troublesome, limping salesman who insisted on displaying his wares. Dad finally dismissed him by firmly telling him to leave. After gathering his goods, the man stomped angrily to his car with a seemingly normal gait, having suddenly forgotten that he had arrived injured.

Mom, Sam, and I actually looked forward to some traveling salesmen, especially those from two companies: Rawleigh and Watkins. They dazzled us with open suitcases displaying an astonishing array of goods. Rawleigh suitcases were the original apothecary shops, offering antiseptic remedies and rubs for man and beast alike: medicated liniments, menthols, ointments, camphor, and salve. And for cooking, Watkins offered colorful tins of cocoa, spices, and extracts. Our eyes bugged at the possibilities. I was entranced with fruit-flavored Watkins Nectar Syrups—so much more elegant than ordinary Kool-Aid. Alas, Mom did not purchase them; always frugal, she bought only no-frill necessities such as ointment and vanilla extract.

A time or two a week, it seemed someone would come up our road simply for a daytime chat. No one, though, could afford to take much time away unless carrying an important message of illness or death or to request help with a specific task, such as baling hay, buzzing wood, or any number of other chores. The preacher from the Methodist church stopped by for visits, calling on members of his congregation. Grandpa and Grandma Walhovd sometimes chugged over from Birchwood, arriving in their ancient Model A. Rob and Inee bumped over to our place in the Green Hornet.

But often we were visited in the evening by one of our many neighbors of Scandinavian descent, primarily Norwegian, like Ole Carlson. The Carlson, Jacobson, Erikson, Mikkelson, Thorson, Stevenson, and Bjelland families had joined Henry Hanson, who had populated the neighborhood with his four adult sons: Martin, Irvin, Art, and Ole. Patriarch John Halberg, long gone, once upon a time had employed men at his sawmill and effectively anchored the Swedish foothold. Not to be outdone by the Norwegians, each of Halberg's five sons—Fred, Ed, Albert, Hank, and Elster—owned a family farm along Highway C. In the 1950s, many voices of Town of Meteor residents still retained the lovely lilt of a foreign accent, even more so in those from the earlier generations.

Coming from an English-German-Swiss family, I often wished that I was Norwegian or Swedish (like most of my friends were). Luckily, Grandpa Walhovd was full Norwegian, so I could claim I was a quarter. When my friends bragged of hearing a few Scandinavian naughty words, I pretended I knew them, too.

Yet I sometimes wondered how we fit in a neighborhood of so many Scandinavians who, despite up to six meals a day, were all skinny. Scandinavian men expected their farmwives to serve breakfast after morning milking and chores, dinner at noon, and then a hearty lunch each day at three o'clock. Farm work halted for up to an hour before men returned to the rest of their afternoon work. Wives served supper at roughly five o'clock before men went out to milk the cows.

After the milking and evening chores, Scandinavian families often dropped in for a visit, whether around the neighborhood or at our home. These guests expected another lunch before departing: sandwiches and jars of home-canned "sauce," meaning peaches, pears, plums, or other fruit. Mom often sent me to get a jar of my choosing.

Topping off the evening lunch, Scandinavians expected a cookie or two served with a cup of coffee. In a pinch, another sweet would do. If Mom didn't happen to have a treat in the house, she would whip up a cake, dancing around the guests in the kitchen. By nine or ten o'clock, one guest or another was sure to say, "Milkin' time comes early."

An evening lunch would be the last of Mom's food preparation on what could have proven a busy day in the kitchen. For no matter what Mom was fixing for the meal, whether dinner or supper, she would extend the food one way or another and set another plate—for neighbors, friends, family, hired workers, or even passing salesmen we took a liking to. Dad would invite all to join us at the kitchen table with a simple, "C'mon up to the house."

OLAF "OLE" CARLSON

Every now and then, Ole or Dad would repeat my favorite story— their first meeting in the summer of 1917, shortly after Dad's family had moved from Indiana. Dad's small log house on Badger Creek stood only a mile or so from Ole's family's homestead. From different directions, their mothers took to the woods with buckets in hand and seven-year-old boys in tow to pick wild raspberries. The season attracted humans and bears alike. So, while Grandma Bessie and Dad wandered through berry patches, they remained wary. When they heard bushes rustling nearby, they were certain it meant a bear. Suddenly, berry-picking Belle and her yellow-haired son appeared, equally startled. Neither mother had known of the others' existence in the neighborhood.

I came to think of Dad's buddy as the Man from Poverty Hill. The windswept high point Ole called Poverty Hill had been his home since he had arrived in the harsh northern wilderness, though he left for a while for a stint in the navy during World War II. He was a man who had never learned to swim, but when the war came, the government sent him on a three-year tour at sea. He recounted the painful experience of being "taught" to swim. Superiors terrified non-swimmers by throwing them into a pool of frigid water. Just as bad, seasickness overtook Ole the instant he boarded a ship and it stayed with him every single day until he again set foot on land.

After the misery of the navy, Ole was drawn to return to, as he called it, "the home place." However, he was never quite sure the

Dad (far left) and Ole (third from left) in second grade at Valley View school

old home place was where he should be; the navy had awakened a taste for the outside world while also providing his living. Plagued by asthma and recurring bouts of ulcers, Ole got by taking short-term jobs now and then and living simply on his veteran's pension. He wrote letters to old navy buddies in his meticulously perfect longhand and fed his recurring urge to get away by taking trips in his pickup to visit his navy friend in the Bitterroot Mountains of Montana. Cowboy country.

Ole's tidy frame farmhouse captured the misty memories of his dead parents. Barney, his dad, was an often-unhappy Scandinavian man. According to Dad, when Barney was old and beyond working age, he would sit at the table looking out the kitchen window. Bitter tears would roll down his cheeks, because he knew his working days were over, and he was unable to think of anything else he could do. After Barney died, Ole would tell of the day that he had finally grown fast enough to outrun the old man, escaping to the woods to get away from the whippings. Was it true, or a yarn made up for the sake of a story? We believed him.

Ole, a man of absolute honesty, was still telling that story when he was an old man.

Belle, Ole's mother, a kind and gracious woman who was by all accounts as sweet as the tinkling sound of her name, had once been a housemother at St. Olaf College in Northfield, Minnesota. She brewed strong Norwegian coffee and made tea in a fine, hand-painted, blue teapot. She served tea in china cups but had the habit of first pouring the steaming liquid into her cup, then on to her matching saucer to cool, sipping daintily as she lifted the saucer to her lips. Ole adored Belle. When Belle breathed her last, Ole walked calmly down our lane and into our kitchen. "I suppose she's in heaven," he said. "If there is such a place." With no further words, he opened the crumpled paper bag clutched in his hands. Reaching inside, he handed my mother the gift he'd taken down from the warming oven above his cook stove—Belle's hand-painted china teapot.

On countless nights, Ole donned a red flannel shirt, or his plaid Woolrich coat, and walked to our home, carrying a bundle of *Field and Stream, Outdoor Life,* and *Look* magazines. He often brought a package of store-bought cookies, or, if he'd just been to town, he brought a carton of ice cream for Sam and me. We treasured his cheery greeting and reports of the latest news, or as he called it, "the scoop"—local folks' farm work, letters from his cousins (Inga and Ingrid), or items from radio broadcasts. He commented upon the weather or what he'd viewed from Poverty Hill. Maybe he'd seen a sow bear and her cubs amble across his field, scared up a covey of partridge while cutting wood, or watched a doe bring her fawns from the edge of his woods to explore his field.

Ole quoted lines of highbrow poetry, laughed when he told Norwegian jokes, and recited lines from silly limericks: "There once was a young gal from Nantucket . . ." (Ole supplied his own child-appropriate endings). We didn't know if his recitations were of his own making, but we did not ask or care, being simply happy for the diversion.

Ole sat at our kitchen table and recounted with Dad their boyhood days at Valley View School, fishing for trout in Badger Creek, and spending days hunting and fishing at Deer Lake. Together they explored the wilderness and the remains of the High Bridge railroad trestle on the logging railroad, left from the white pine harvest. Like Tom Sawyer and Huck Finn, Ole and Dad told stories about their river adventures, only theirs were on the Couderay and Chippewa. While fishing for muskie once, the canoe floated against a partially submerged log. Suddenly, the "log" thrashed wildly under the canoe while the biggest muskie the boys ever saw swam safely away.

The pair reminisced about folks long gone, and Ole regaled us with tales from lumberjack days at Camp 5, Swan Creek, and Poppyville: "I remember the time yer uncle Jerry raided Cookee's cupboard an' found himself a cake. Gobbled up haf it."

"Hey there, that's one gud raisin cake ya baked there, Cookee," Jerry said.

"There ain't no raisins in that cake . . . them're a nice croppa ants!"

"Yah, yah," Ole would say when the conversation stalled, biding his time until another idea came to mind. When there were no more, Ole teasingly mimicked talk from WCCO radio as Mom tended to the perking coffee pot, "Ya got any a them horses doovers there, fer lunch?" No matter what the time of night, Mom served the hors d'oeuvres "lunch" of sandwiches, fruit, cake or cookies, and strongly brewed coffee.

At ten o'clock, Ole would pause as he stepped onto our porch, never failing to scan the sparkling sky and point out visible planets and the stars he'd read about in his current *Astronomy* magazine. Then he stooped to tightly pull the laces on his L. L. Bean boots, gathered a bundle of our *Alaska Sportsman* magazines in return

for those he had brought, and, boots crunching gravel, strode off
into the starry night, back to his bachelor existence.

UNCLE JERRY

Jerry and Dad often worked together and got along most of the
time, despite their very different personalities. Dad was quiet
and reserved, like Grandpa Sam—a gentleman who dressed in a
proper three-piece suit when visiting on Sunday. Grandma Bessie,
though opinionated and outspoken, remained primly true to her
English upbringing. Gerald—as she called her firstborn son—
often mortified her. Living across the road from Jerry, his wife,

Millie, and their daughter,
Shirley, Grandma Bessie
had occasion to witness
Jerry's daily behavior. As
did others in the immediate
neighborhood.

One day while Jerry
was on the ground pick-
ing rocks with Millie, their
horses became so fright-
ened by Jerry's roaring that
they bolted and ran. Jerry
and Millie ran helplessly
after the wagon as it jolted
away, horses tearing across
the rough field with seven-
year-old Shirley aboard. She
clung on for dear life amid
rolling rocks. Neighbors
heard the frantic shouting
and Jerry's obscenities and

Uncle Jerry on the Chippewa River

rushed to help. Fortunately, the horses and wagon finally drew
to a stop when they reached the barbed-wire fence at the edge of
the field. Jerry's bellowing often befuddled his animals, as well as
his neighbors and his family.

Millie had enough spunk in her to counter Jerry's outbursts.
She was fun-loving, always ready for a card party, and eager to
socialize. Millie worked alongside Jerry every day, kept him on
task, and did plenty of the hard work herself. Instead of the usual
dairy and beef cattle operations of the day, Jerry boldly chose a
triad of new ventures: raising experimental sheep for wool, keeping an aviary of bees, and planting an apple orchard on terraces
he'd created on the hillside—using the forefront of agricultural
practice. Jerry was the first in the neighborhood to take advantage
of learning terracing from experts at the University of Wisconsin
in Madison. UW–Madison sheep researchers based in Spooner
also followed his pure-bred flock of Hampshire sheep and tested
their success in northern climates; they were not known as a particularly hardy breed. In addition, Jerry was active in sheep-raising
and beekeeping organizations.

Jerry's sheep pasture and hives were adjacent to our home.
In springtime, the bees pollinated Jerry's apple orchard, and in

Uncle Jerry's Hampshire sheep

early summer, Jerry took his hives to pollinate the blossoms in cranberry bogs twenty miles away, turning out delicate cranberry honey.

Shirley and I often played on a giant boulder near the hives while her parents worked the bees. Donned in protective white suits and veiled hats, they added supers to build the hives taller and removed the flats to their small honey house where they extracted the stored honey from the boxy frames. Then Jerry and Millie spun the honey from the comb and bottled it, always leaving our family with a square of honeycomb.

The sheep impacted our home, too. When Jerry and Millie drove their flock up the tote road along Badger Creek, Jerry's voice carried as he shouted all the way. On summer evenings from our front porch, we watched rams butting heads and ewes watching over their lambs at play, bouncing into the air and playing leapfrog. The distinctive smell of sheep manure wafted through the cool night air. Just beyond our house, Jerry corralled his sheep before driving them one by one into a narrow chute leading to a deep vat containing sheep dip: a foul-smelling concoction that combated a disease caused by parasitic liver flukes. The sheep would emerge with wool coats thoroughly drenched and, for a time, dyed a rich purplish blue, but not before resisting with a struggle. The sheep baahed and blatted while Jerry cursed, but he finally forced each animal into the vat.

Jerry expanded my vocabulary; whether on our farm or on the streets of Exeland, it was from him I learned words not heard in my own home (unless Dad hit his thumb with a hammer). His speech was often punctuated with profanity, just as an extension of his everyday expression. He seemed to give no thought that anyone would be offended. No one challenged his use of words. If they had, he may well have offered a sincere apology, unaware that his next response continued with the usual profanity.

Jerry caused a stir everywhere he went, and he needed no

introduction; his big frame and booming voice announced his appearance. He voiced opposition to plans of any kind, blustered at meetings, and loudly relayed his latest adventure on Saturday nights in Exeland. He kept a bit of exaggeration in his stories and enjoyed engaging in argument and provoking anyone to spar with him over any issue. His lack of formal education did not keep him from vigorous discussion. Once, at the end of one of Jerry's rants, I heard a bystander snicker when Jerry concluded, "And I don't like yer *altitude*, either."

But Jerry held no grudges and walked away as if it were simply his entertainment. At the same time, he could display a soft heart and express great sympathy at the plight of a poor soul. Once, Jerry had even rushed to town and rallied Harry Taylor to rescue a bull belonging to the area's only Polish immigrant.

Those logging days no doubt contributed to Jerry's personality. Presumably that was where he picked up his salty language. Once, working in the woods, he'd chopped his foot with an axe. Lumberjacks told him to leave; he was getting blood poisoning. As he walked along the rough road to find a doctor, a car stopped. The driver offered to take Jerry to the doctor—but only for what was a lot of money at the time. "Go ta hell!" Jerry retorted. "I'll walk!"

Despite Jerry's rough-and-tumble young life and spelling a word differently each time he wrote it, Jerry was bright. Later in his life, regretting his folly of "havin' no truck" with education, he encouraged Shirley to seek all that she could. He realized the deficit it had created. Still, his natural smarts, salesmanship, and entrepreneurial bent had provided a good living. Jerry was always one to take advantage of every economic silver lining.

Jerry's ventures even tested the wildlife of northern Wisconsin. Toward fall, Jerry's livelihood proved too much temptation for the black bear population. Like Goldilocks, the time was just right to flatten his corn crop, rip into ripe ears, and tear them from stalks. Crisp, sweet apples drooping from branches were just right for munching. Docile sheep and lambs were just right for a meaty main course. The sheep, being sheep, were helpless against the marauding bears. Even then, the bears weren't done, saving for last the most delicious treat of all: sweet, sticky honey. Bears battered and smashed hives that had been dripping with the fall crop. Bees buzzed in fury while bears fled, but the rich honey was gone.

To stem the loss, the Wisconsin Conservation Department gave Jerry permission to trap and hunt the destructive bears. Once bears had located a flock of sheep, especially one surrounded by corn, apples, and honey, everyone knew they would return. If left unchallenged, bears could destroy an entire flock.

Jerry then offered yet another item for his customers, proclaiming its benefits: "Bear meat, better 'n beef! Sweet tastin'—'n' ain't it damn good for ya too!?"

INDIANA RELATIVES

Though Mom might be weary and Dad could be cranky, summertime's work had to be done in a timely fashion. In our world, *Ya gotta make hay when the sun shines* was more than an aphorism. And amid the harried summer scene, one occurrence repeated without fail. Although a welcome diversion, it also complicated our lives.

An incoming postcard informed us that our Indiana relatives would arrive at the depot in Exeland on the evening train. The company would need someone to run to town for a pick-up. Somehow those postcards seemed to come just in time for us to scramble and arrive smiling at the depot. But if they were driving, the postcard arriving in our afternoon mail often read:

Driving and leaving Indiana this morning. Will stop in Ladysmith. Plan to arrive in the north woods by tomorrow afternoon. Love to Uncle Sam, Aunt Bess and All. See you soon.

Aunt Effie, Cousin Gladys, Paul and the Boys

The cryptic news presented a problem. When was "this morning" or "tomorrow afternoon"? How much time did we have to
get ready? Since we had no telephone, it made sense for the relatives to send a postcard. But maybe they could have mailed it
a bit sooner? Or included dates? We pictured them posting the
missive as they backed their car out of their Indiana driveway and
hit the road north.

Thus, a postcard proclaiming their impending arrival sometimes reached our mailbox *after* our relatives had already made
their appearance in Grandpa Sam and Grandma Bessie's driveway.
The silver lining? No time to worry about preparation. For when
the message preceded company, it threw the three Prilaman families into an instant tizzy. Clean the house. Do as many chores as
possible. What to cook? Where would they stay? How would we
entertain them? Who all were they bringing *this* time?

The relatives, Mom never failed to point out, were Dad's "shirt-
tail relatives," not closely related. Mom often grumbled that *her*
relatives were less intrusive—and always welcome. "Some of
those Indiana relatives like to put on airs," she'd mumble.

Soon enough, Aunt Effie, Cousin Gladys, Paul, and the Boys
spilled out of the car, happy to have arrived after the long ride in
anticipation of their wilderness retreat. They stretched, shouted
hellos, and took in great gulps of cool country air. Kids rushed
off for a look at Badger Creek and to hop over the rocks. Mom,
Grandma Bessie, and Millie rushed about making everyone welcome; they put together a meal suitable for guests whose hun-
ger was already fueled by the fresh air. Sam, Shirley, and I hung
around to watch the action down at the creek. Inevitably, the

novice rock hoppers would sooner or later tumble into the drink, an event we did not intend to miss.

Our guests were fun-loving, kind, and generous folks who came bearing small gifts: a kaleidoscope for me, a tiny birch bark canoe and teepee for Sam, and once, giant yellow and red lollipops on a stick for each of us. We didn't dislike them, though it was obvious that we lived in entirely different worlds. Our world wasn't even small-town America—simply a small farm on a narrow gravel road through the wooded hills, leading to barns and pastures littered with wet, warm, squishy cow pies—something every farm kid has stepped in. They, however, lived in a town about eighty miles from Chicago, spanning the flatland of corn country.

While our company gushed over the charm of the Northwoods, the crisp piney air, and the quaint habits of country folk, we set aside some of our crushing summer chores and attempted to entertain and be social and polite to our kin. Still, carrying on with the necessary chores tested everyone's patience. While Grandma Bessie whipped up a cake or a pan of biscuits, the Boys went down the path to fetch water from the spring in Badger Creek, my grandparents' only source of water. A novel chore for city folk, they managed to slosh back with their pant legs dripping and a half-bucket full.

Even so, Grandma Bessie was especially pleased to have some of the homefolks around to speak of old times. The women chattered happily in her small kitchen or as they strolled out to see the gladiolas growing in her garden, pull a batch of carrots, or pick string beans for supper. They discussed the latest dress they had sewn, a needlework project, and the theme of the preacher's last sermon. All the while, I tagged along, listening in, mesmerized by the cadence of their voices, so different from our nasal Midwestern tones.

"Ah's half-timpted ta give the preachah a piece of mah mahnd, after he wen' on fer more 'n an hour las' Sunday," Aunt Effie once declared.

One voice or another nattered on in language foreign to our ears, discussing topics we considered inane. It took great restraint to contain our eye rolling as Aunt Effie regaled everyone with tales of the most recent Ladies Aid Luncheon or the delectable dessert served last Wednesday by her bridge club. They met on a bridge? We silently wondered about their very different lives.

We, who took mosquitoes for granted and were riddled with bites all summer, thought it ridiculous that Aunt Effie could fuss endlessly over a single itch. Sam and I hid our snickers each time it became the topic of conversation but howled with laughter in private.

We were more likely to retell stories of the bees' nest in the raspberry patch, the day the cows got out of the fence, or the messy trials of teaching a slurping newborn calf to drink by dipping his head into a sloshing bucket of milk. The questions our guests asked amused us—all about farm and country things we took for granted. Had a skunk ever sprayed us? Did we see black bears? What did a porcupine look like? Could it really shoot its quills? Did we hear wolves howling at night? And, what is *that*!?—when we "failed" to avoid the pine snake sunning on a rock pile as we escorted our cousins around the farm.

That was a variation of a black rat snake, easily four or five feet long and looking like a rattler, though pine snakes were harmless. They were also fond of stretching out in the sun around fallen pine logs and hanging from the rafters of aged log buildings. They were remarkable climbers whose presence we came to expect, but they could still startle—even Queenie and Joe. One day Dad returned our horses to the barn to remove their harness and they reared in fright: a long snakeskin dangled from above the stall. Dad figured the snake had moved on after shedding its skin while slithering across the rafters. Like the horses, I wasn't fond of pine snakes either, but they came in handy when it came to educating our cousins.

So did the electric fence. Dad put up an electrified wire (powered by a low voltage battery) to contain the cows in the pasture. Touching the fence resulted in a shock, but we liked to experiment. A piece of dry grass barely grazing the wire resulted in a small shock; straw pressed against the wire increased the jolt; a green stalk, containing moisture, gave a satisfying *zap!* Sam and I dared one another to prove our courage. We used this information to our advantage when showing our cousins the ropes of farm operation.

Sam and I also invited our city cousins to ride the growing calves in the confines of the smelly pen. (Though Dad frowned on the practice, we got away with it occasionally. But not the time we fell into reeking piles of manure with our Janesville cousin, Don. Everybody knew darn well what we'd been up to.) The Boys declined that offer, but Sam successfully persuaded, "Ya gotta try riding a *cow*, then." As if on cue, Brownie promptly headed for the nearest low-hanging branch and harmlessly swept off her rider onto the pasture. Sam and I snickered; the Boys'd had enough farm recreation for a while.

One evening Mom got into the spirit, entertaining our gape-mouthed visitors with a tall tale of Paul Bunyan's lumber camp at our own Deer Lake. She said many guests had suddenly appeared, so Paul sent some men out on a hunting party and had the rest cut an acre or two of trees, to fuel gigantic bonfires flaming all around the shoreline. Paul's crew rounded up and skinned deer, rabbits, raccoons, and squirrels, and dropped them into the steaming cauldron—Deer Lake itself. They threw in a skunk or two for extra flavor. The lake began to boil and hiss, steamy as a tea kettle. Babe the Blue Ox hauled a sled load of rutabagas, two loads of potatoes, and another of onions; the lumberjacks tossed them in, too. It wasn't long before the feast fit for Paul and his guests was served. Bears came from miles around to gaze in wonder and slurp up gulps of the tasty stew. By morning, the bonfires had smoldered out, the lake had cooled, and when the guests awoke from their pine needle beds, Paul sent them on their way. The skeletal carcasses of deer washed up on the shore were licked clean by the bears before they left for the woods.

Mom then sent the Boys off to bed with wishes of sweet dreams, but as they climbed the stairs to my grandparents' attic, they heard the eerie call of hoot owls—and baying and yipping coyotes. Presumably, the Boys lay awake for hours wondering about the stew Grandma Bessie had served them for supper.

Teasing our cousins provided no end of amusement, as did watching them amuse themselves in our environment. Cousin Gladys, an accomplished amateur photographer, incredulously found subjects for photos in our ordinary farm life. Brownie became the star of her very own nine- by twelve-inch portrait. A drive through the Lac Courte Oreilles Indian Reservation twenty miles away, to visit a St. Francis Mission School and church in the community of Reserve, also supplied material for Cousin Gladys's black-and-white photos.

We were as elated as our guests when we took rare trips away with them on a summer's day when there was work to be done: floating in Dad's forest green canoe on the Chippewa River, visiting remote Deer Lake, or having a rustic picnic along the brushy banks of the Couderay River, not far from burial sites on the reservation. We could still see the Native American traditional wooden-gabled grave coverings. But the highlight of summer vacation in the Northwoods was the year a family of black bears purposefully came to harvest Jerry's orchard, cornfield, and hives.

When the excitement finally died down a few days later, our company's car drove slowly out of view through a cloud of dust rising from the gravel road. When that southbound car rounded the last bend, an audible sigh of relief could be heard in simultaneous exhale. Relief was short-lived—for now we missed their entertaining presence. Back once more to the drudgery of summer chores, now with even more urgency for the time we'd missed.

But perhaps we did the chores more happily, breathing in that cool country air, in the quiet peacefulness of our quaint countryside.

ROB AND INEE

Everyone loved visiting Rob and Inee. Friends, neighbors, and strangers alike were made instantly welcome in their home. Rob would emerge from his workshop to greet guests, puffing on the pipe loosely clenched in his teeth. He wore a faded and neatly patched flannel shirt and pants held up by wide, yellow Brace brand suspenders. In winter, and much of the summer too, he wore long underwear tattered at the collar and warm wool pants; a red paisley bandana trailed from his pocket. His husky frame told of Inee's good home cooking. Rob was jovial, soft-spoken, and congenial. I never heard him speak a harsh word against anyone. He seemed happiest when sharing a long-forgotten skill with a child, joking around, or showing us how to whittle.

Kids loved exploring the creek and the yard, and we ran to the hand-operated well to pump it until cold water poured from the spout. A dipper hung ready for use; there was lovely simplicity in taking a refreshing drink from that long-handled enamel dipper.

I think often of Rob and Inee's little house beside the road, the faint light of gas and kerosene lamps shining through the windows. After darkness fell in the warmth of that room, they sat in wooden rocking chairs, reading by dim, flickering lamplight; Rob believed electricity was unnecessary, proclaiming, "Them bright lights'll ruin yer eyes."

To visit Rob and Inee's was to step into a time capsule, to

receive lessons in reusing and "making do." In their daily routine and in all kinds of weather, Inee carried buckets of water from the pump, Rob carried in wood for the stoves, and each of them followed the narrow path to the outhouse. One day Inee remarked, "I somehow threw my back outta kilter." Later, painfully inching her way toward the outhouse along an icy winter path, she slipped and fell flat. When she rose carefully to her feet, she realized the tumble had miraculously realigned her back and the pain was gone.

Rob and Inee lived resourcefully and maintained a huge garden planted according to advice offered by *The Old Farmer's Almanac,* which hung from a string on a nail in their kitchen. They

From left to right: Rob, Aunt Effie, and Inee, 1956

planted strictly by the phases of the moon—potatoes *always* on Good Friday. Inee famously grew a special heritage bean, known locally as "Inee beans." The tall trailing beans grew on poles; and she picked, dried, and shelled them in late fall. Then she soaked the beans in water, placed them in a brown bean pot, and made sweet, steaming baked beans. She also canned sucker fish in the spring and meat in the fall: chickens, game birds, and venison.

As the ultimate handyman, Rob had cut small trees for poles from his woods and placed the poles vertically to build the outhouse, garage, and a small barn. When Rob and Inee visited the Jacobsons' house across the pasture, they were aided by a wooden stile, allowing people but not animals to step over the barbed-wire fence. Rob had built it to prevent ripping out the seat of his britches.

For as long as I knew him, Rob never held an ordinary job. He often recited Matthew 6:28: "Consider the lilies of the field, how they grow; they neither toil nor spin." He expected things to work out for the best, and yet, both he and Inee evidently believed that very hard work was still necessary. At one time, they had made their home in a canvas tent for months, surviving on ingenuity, wild game, fish, and Rob's trap line, set to catch mink, muskrat, beaver, raccoon, and fox. He'd sold the pelts to a buyer who came around every now and then and paid a bounty for each: less than a dollar for a muskrat, up to ten or twenty dollars for a high-quality beaver pelt. Sometimes he'd say, with sadness tinging his deep voice, "Gol dern, when I was a boy, flocks of sandhill 'n' even whoopers gathered in fall; not many of them're left, no more."

Rob worked for short periods of time as the Town of Meteor Assessor and was known as a man who dispensed fair assessments. His easygoing personality, friendliness, and honesty made the job a perfect match for him. He also added to his World War I pension, their main source of income, by charging a small fee for leather work, woodwork, shoe repair, and other odd jobs.

Rob's workshop beside the creek seemed his place of refuge. Kids and adults alike watched in wonder while he worked. He carved lengths of wood into durable canes, whittled plugs of wood into effective muskie baits, and mounted countless racks of deer antlers for display. Rob could stuff a bird or mammal using his taxidermy skills; lace and bend a pair of snowshoes; shape a pair of skis. With a heavy-duty sewing machine, he sewed canvas and leather game bags, made scores of leather mittens to wear over woolen ones, mended leaky tents, and made leather belts, straps, and pack sacks. He could equally well fix a leather harness or repair baseballs and mitts.

As an accomplished cobbler, Rob had a steady stream of customers. Every kid in the neighborhood wore shoes or boots with Rob's work holding them together, whether it was new heels or soles. He even created shoes, moccasins, and slippers from his

Rob in his workshop

deer kills. He had learned how the Lac Courte Oreilles used a fresh deer brain to tan its hide into soft, supple deerskin.

While he worked, sometimes with a pipe in his mouth, Rob was a wealth of folk wisdom and good humor. His customers came as much for a good visit as a repair. He chuckled as he asked, "Got rheumatism? Try beatin' yer legs with thistles." (Puff, puff.) "Won't help the rheumatism, but ya won't complain of *that* pain!"

If Rob's workshop beckoned invitingly, Inee worked her own magic inside their house. She displayed her own brand of welcome when a carload of visitors drove into the yard. As the screen door slammed behind her, she'd appear with a shy but happy smile while wiping her hands on the bibbed apron always tied around her waist. She dressed in a cotton housedress, with dark-colored rayon stockings sagging around her thick ankles, and she wore her braided gray hair wound around her head. With no children of her own, she was always pleased when kids came. She invited passersby to stay and visit and felt insulted if the guests did not remain to dine for the next meal—then stay even longer to gab on through the evening and partake of a late "lunch" too.

Thus, no one ever left without sharing a fine meal of home-grown food, and Inee could whip it up in minutes. She'd disappear out the back door long enough to retrieve "fixin's" from the cellar, which she entered through a slanted door and descending stairwell. The cellar's cool, hard-packed dirt floor was their only refrigeration in summer. Its wooden shelves held Inee's canned fruits and vegetables of every kind. Root vegetables and dried apples rested beside a large potato bin. After gathering an armload, and with a slam of the screen door, Inee would return and pick up the conversation with company.

Inee managed her kitchen with amazing efficiency and skillfully put together the meal using her always-warm wood-burning cook stove with a warming-oven above. Hovering at the Hoosier—a freestanding kitchen cabinet that included a storage bin for flour and a small work counter—she went to work. While visitors gathered around a tipsy wooden table covered in faded oilcloth, she retrieved sugar and flour from the cabinet and stirred up a cake for dessert.

Those who knew Inee laughed at one peculiar habit but considered it a fair trade. After her company had enjoyed a meal and offered to help with the washing up, she'd reach deep into her cupboards and extract pan after pan of dirty dishes. On a small table against the wall stood dishpans to wash and rinse. Guests took warm water from a reservoir on the side of the wood stove and dipped cold water from a bucket. Still chatting, everyone helped her wash, dry, and put away nearly every dish she owned.

Inee was prized for her knowledge, and her friends and neighbors often sought her advice. She knew how to make anything, from cleaning solutions to doughnuts. Her bread, rolls, cakes, pies, cookies, and handmade pastel mints were well known. Her rich, dark chocolates, filled with mashed potato and coconut, were special treats any time of the year (and were Sam's favorite).

In the summertime, Inee made batches of root beer for guests. She purchased a small medicine-like bottle of Hires Root Beer Extract, available at Veness's, and a cake of yeast for fermentation, like that used in bread making. Her recipe: add water to extract and yeast, pour into ordinary sealed canning jars, and let sit for a week or two. Once, she placed the jars out of the way to ferment, under a bench behind the kitchen table. That's when the preacher came to visit. After the usual conversation and prayer, it was time for afternoon coffee. He took his place behind the table just as a jar of the root beer exploded. Inee said the preacher laughed nervously, but she calmed him with a piece of pineapple cake and coffee. Although initially she had been horrified, she always told the story with snorts of laughter and remained certain that the preacher believed she brewed beer, made wine, and distilled moonshine, too.

In or out of the kitchen, Inee was a woman of many talents. As a proficient seamstress, she often pumped the foot pedal on her treadle sewing machine to mend and patch everything from overalls to wool pants to shirts. She knit woolen inner mittens to wear inside Rob's deerskin outer mitts, darned his socks, and stitched up holes in his long underwear. She crocheted and knit gifts for babies and brides alike, producing cross-stitch aprons and needlework for special occasions.

Whether at home, visiting, or traversing the dusty roads to town in the Green Hornet, Rob and Inee always stayed together—like a set of salt and pepper shakers. And like a Boy Scout, Inee was always well prepared. Even if she left home in early morning, she carried with her a flashlight so long that it required multiple D batteries. She took it everywhere and checked it routinely to be certain it still worked. One would have thought she would be used to the dark; she'd lived without electricity for her entire life. But the flashlight was a permanent fixture. She also carried a bulging purse under her arm—an always-tightly-clutched purse—as if at any moment it would be torn from her grasp. It functioned as a file cabinet and was jammed with carefully organized wads of information, clippings, and coupons cut from newspapers and magazines. If anyone asked when trout season began, when Easter came this year, or when to expect the next full moon, Inee was on it. She'd extract a clump bound by a rubber band, locate the proper clipping, and produce an answer.

Rob weathered Inee's eccentricities, ups and downs, and scoldings for infractions, taking his wife and her ways in stride. For while he remained even-tempered and unshaken even at the most startling pronouncement of news, such as a house or barn fire—his response never more riled than, "Well, I'll be gol derned"—Inee sometimes displayed a sharp and curious edge to her personality. Even those who knew her well became puzzled when she chose to snub someone. It came on suddenly, could last for months at a time, and had no apparent reason. When this happened, someone would say, "I dunno what happened, but Inee's got a mad on."

I especially think of Rob and Inee each fall, since Rob and I shared an October birthday. He'd often say, "There's no better way to celebrate a birthday than under a full harvest moon . . . and a slab of Inee's pumpkin pie."

Rob and Inee were also empathetic to those in distress or grief and were the first to arrive with food and provide solace by just sitting with the ones who needed it. They would also open their house to anyone in need. One time, Richard Halberg took a near-fatal tumble off of his new bike, right in front of Rob and Inee's house. Richard's older cousin Eric had then picked up the silent, still, and bleeding boy and run him into the house, where he lay unconscious on Rob and Inee's bed. In World War I, Rob had seen his share of trauma. No one has ever known just what Rob did then, but he may have saved the boy's life. Richard was then transferred to St. Mary's Hospital, where he went into a coma. Eventually, he recovered, and two months later, he returned to school.

FRANK THE MILKMAN

Promptly each morning, a boxy truck rumbled down our dead-end road, spitting gravel as it slowed and turned into the spur of the driveway. Frank, the milk-truck driver, always arrived by seven o'clock. Day in and day out, year-round, Frank was not to be kept from his appointed rounds, which started at dawn. He drove the Ladysmith Co-op Creamery truck and exchanged empty, clanking ten-gallon cream cans for cans full of milk.

The creamery identified milk cans with a farm number— ours was X-130—painted in black on the lids and side, then paid farmers for the amount and quality of the milk. Frank brought the monthly milk check as well as the butter and yellow cheddar cheese we ordered. When other farmers updated, replacing cream cans with bulk storage tanks for Grade A milk, Dad decided our farm would remain small; we did not install the pricey cooling systems.

Frank's unrefrigerated truck meant milk must get to the creamery quickly in summer. Like Frank's, Dad's morning started early. He awoke to the brassy clattering of our alarm clock at five o'clock, donned his gear, and headed out the door—work before breakfast. Once in the barn, Dad rushed to complete the milking and get the cans of milk into the cooling tank for a quick, albeit minimal, cool down before Frank arrived.

Frank swung the heavy stainless-steel cans from his truck smoothly, one in each hand, to the platform Dad had built on the hillside. From the cooling tank, Frank lifted the full cans, each weighing more than a hundred pounds. He was not a big man, but he hoisted them by their earlike handles with ease and quickly settled them into racks on the truck. With rugged efficiency, he wrestled everything into place and was soon ready to depart.

Frank was a quiet man, not quick to engage in conversation that would delay him, nor was he prone to gossip. However, in those pre-telephone days, perhaps he thought it his duty to deliver news from town or from along his route. The death of an old-timer, a tragic accident, or a misfortune—he knew we'd want to know. Frank brought news of startling interest when I was in sixth grade. On Christmas Day, my school had caught fire in the night and burned to the ground.

Getting the milk to market sometimes presented challenges. Springtime, especially so. During the thaw when frost began to rise, sticky mud slowly oozed to the surface and made the roads as springy as a trampoline. Vehicle weight limits were imposed until sun and warmth dried up the rural roads. In a cold, wet spring, roads could be impassable for several weeks at a time.

But the milkman must go through. Knowing the muddy conditions happened most years, local folks were prepared for this season. As long as he could, Frank rode the gravel road ridges, perching his truck above the mud. If the truck slipped in and became mired, a farmer with a tractor pulled him out. Then the men clearly knew the road had become impassable, and they would have to haul their milk with tractor and wagon, or even horses, to where Frank could safely drive his truck. Farmers pulled drags over the road to level it; the township sent the road grader through to bring the mud to the surface and dumped loads of gravel in particularly bad holes, until the roads finally went back to normal.

Winter presented a different challenge because of frequent and heavy snowfall, ice, or drifting and blowing snow. Frank's truck took on a dual purpose when he mounted a snowplow on the front, and his was always the first vehicle to cut its way through to our home. After clearing the main roads, a Town of Meteor snowplow would arrive to fully clear our half-mile, dead-end road.

Nothing short of a massive blizzard kept Frank from his route. He was also the first to report a school bus stuck on the road, broken down, or in a ditch. On occasion, he even took on the role of school bus driver by transporting stranded teenagers to Exeland so they could board another bus going to Bruce High School.

Frank was always kind and helpful, but given the limited maneuvering space for his milk truck, an occasional mishap was bound to happen. One Thanksgiving weekend, when my friend Sonja Halberg and I were teenagers, we devised and implemented a moneymaking scheme. We first combed our local woods and gathered bundles of dainty, green Princess Pine. Then we bent wire coat hangers into circles and wrapped and attached the piney strands to create decorations for the Christmas season. By the end of that long weekend, we'd worked tirelessly to produce a stack of ten or twelve perfectly round wreaths, all with perky red bows.

Stepping back to admire our work, we were already dreaming of the Christmas spending money we'd earn. We piled our beautiful garlands on top of a tall snowbank at Sonja's house for safekeeping until we could put them up for sale in Veness's mercantile store in Exeland.

On Monday morning, after a sleepover at Sonja's, we were running a bit late when the school bus and Frank arrived at the very same time. Frank would have to turn his truck around to get out of the way, so we sent Sonja's brother, Johnny, to tell the bus driver we'd be out soon—and to warn Frank about our wreaths. In Johnny's haste to board, he forgot. We raced out just in time to watch Frank's big double wheels grind over our wreaths and smash them into flat pancakes. We didn't salvage even one single wreath or earn a cent for our expense or efforts.

It sure cut into my Christmas giving plan.

PICNICS, PARTIES, SHOWERS, AND SHIVAREES

Early in the summer of 1953, the winter's hay was gone from the haymow and the new crop was not yet in—in other words, perfect timing for a party in the barn. Jean and Luke Birdsill and Valley View 4-H Club members swept out the last of the straw and dust, brewed up batches of homemade root beer, and popped mountains of popcorn to sell. Jean engaged the Roy Woods Band of Ladysmith to provide the music. On a mild and dry Saturday night, people arrived early from miles around and parked their cars in a rough-mown hay field. Folks danced until the wee hours of the morning, and the successful event was fondly and long remembered as Birdsill's Barn Dance. However, it was the one and only dance of its kind, as people realized it had presented a real fire hazard—despite the barn's apparent cleanliness. In those days, men enjoyed the evening by lighting pipes, rolling cigarettes, and smoking happily in the highly flammable barn.

Most of the rural gatherings organized by farmwives in the summer held no such danger. There were picnics on each holiday: Decoration Day, Fourth of July, and Labor Day. Families brought their own tableware, plates, and cups in a picnic basket or box. If tables were in short supply, sawhorses supported boards to hold the feast. Men arranged seating by placing planks over round chunks of wood brought from the woodshed, then women piled tables high with the usual potluck fare: assorted hot dishes and sandwiches, which now had the addition of summer's newly made dill pickle spears, sliced sweet pickles, and potato salad. As always, coconut cakes, marble cakes, and yellow cakes arrived in flat metal pans. But the best was in a green metal cake carrier: Verna Hanson's mother, Flossie, always brought her sticky three-layer cake with snowy-white frosting.

Year round, women planned and hosted parties for special events: housewarmings, anniversaries, birthdays. When Verna's grandmother's house burned to the ground, the community threw a humdinger of a fire shower for Ethel Ploger. Wedding and baby showers were held after evening chores and milking so that whole families could attend. Only two families owned crank phones in the neighborhood, so word arrived in the mail with a postcard.

In our neighborhood, we celebrated weddings with a shivaree. Traditionally, folks held this wedding night prank of clanging pots and pans to surprise and serenade the couple soon after they had gone to bed. The mischief-makers expected the groom to invite them in for late night partying and refreshments.

Mom and Millie often recounted one shivaree from their pre-kid married lives. Mom, Dad, Jerry, Millie, and Ed and Daisy Halberg had set out on foot following a rough logging trail through the woods. They left late in the evening and planned to arrive unannounced at the log cabin of recently married Lawrence and

Florence Bjelland. Knowing the newlyweds would be unable to provide libation, they rolled a keg of beer along with them. After dark, they pounded on the door and further announced their arrival with ear-splitting clangs produced by hammering the giant buzz-saw blade conveniently located just outside. The couple leaped from bed, sheepishly greeted their guests, and invited them in. Their supplies were meager; however, the hindquarters of a just-butchered deer still lay on the kitchen table, and as the night wore on, the partiers consumed the keg and grew hungry. Lawrence cleaved the venison into slabs, Florence fried it up in a cast iron skillet, and they all feasted. When it was time to trek back home through the woods, Mom and Millie, who were the most clear-headed, led the way. They hoisted a kerosene lantern high, lighting the trail for the happy-but-bleary-eyed revelers, and arrived home just before dawn—in time to begin the morning milking and chores.

When I was small, our neighborhood held a more civilized shivaree: we banged pots and pans in honor of the newlyweds only some length of time after a couple had been married and not at their home. Verna's grandmother and mother often organized it and notified every family in the neighborhood. Soon, even that tradition gave way to just partying with a keg of beer in the garage or milk house at a hosting farm. Ed and Daisy's large home—originally John Halberg's home and sawmill boarding house—was well-suited for a crowd. And it had an attached garage for dancing. Records and an occasional accordion provided schottische and polka music for the evening. Kids played games like kick-the-can in the mysterious after-dark or climbed up the sawdust mountain out back where the sawmill used to be. Late in the evening, eating brought the fetes to a close, but sometimes they carried on well into the night. No matter how late, for most partiers, Bossy and her cohorts would still need attention at dawn.

METEOR COMMUNITY POLITICS

Our Sawyer County library and the county seat were nearly forty miles away in Hayward; like most people in our neighborhood, we traveled there rarely, only to go to the county fair or the courthouse. Still, folks made the effort to be knowledgeable, using the tools at hand to educate themselves. The Wisconsin free traveling library provided books, and some families in the neighborhood even owned a set of World Book Encyclopedias. Neighbors often had visitors stop by to look up desired information. Families were consumers of news and entertainment radio broadcasts. Mail delivery supplied homes with ample reading material, including magazines of every kind, such as the one alongside Dad's stuffed chair, the monthly black-and-white *Wisconsin Conservation Bulletin* (it sat next to the Bible). Dad read the bulletin cover to cover after Rob did, and on their next visit, they were sure to debate the latest articles, comparing their own observations of the natural world with what they'd read in the publication. Families often exchanged news magazines, as we did with Rob and Ole.

Local and national daily newspapers arrived in the mail as well. And, for years, Elsie Zesiger reported our weekly local news in the Meteor Hills News column for the *Ladysmith News*. It offered titillating neighborhood society bits and bobs, including who had visited whom, who had gone to the doctor, who had attended someone's piano recital, and who had hosted the most recent Ladies Aid luncheon. Elsie included observations on the progress of farm crops, seasonal phenology, and the contentment and challenges of country living.

In contrast, the national papers provided a broader view of the outside world and opinions, political and otherwise. Dad regularly read Drew Pearson's column, Washington Merry-Go-Round, a syndication sometimes taking public officials to task. Our family also received monthly newsletters from our state and national representatives. Dad sometimes contacted our representatives on pressing matters, such as statewide school consolidation.

For many people, education had been limited. Few, if any, of our neighbors were college educated, though teachers held normal school certification. Some neighborhood women had gone to live in town when they reached high school age. Because there was no public transportation, girls worked for the family who boarded them while they earned a high school diploma. For many men, like Dad, formal education had ended at eighth-grade graduation, when they went to work on the farm or logging in the woods, as Jerry had. The basic knowledge they learned by the end of eighth grade they then applied along with their vast and practical horse sense, finding solutions to myriad everyday problems. They became competent in weather prediction, carpentry, finance, animal husbandry, agronomy, medicine, time management, logistics, and a host of other necessary skills.

Grandpa Sam atypically held a teaching degree from the Normal School in Valparaiso. Both Grandma Bessie and Grandma Walhovd had rural teaching certificates and gave piano lessons. Grandpa Walhovd worked with his hands. Remarkably, Mom and each of her seven siblings had earned diplomas at Birchwood High School.

When it came to politics, people honored the privacy of the ballot and opinion. As a matter of etiquette, they did not ask others how they voted, and political discussions took place behind closed doors. Of course, Jerry was a notable exception. He delighted in publicly arguing politics with his friend and neighbor, Luke. Each relished the lively exchange—openly and loudly discussing their differences on the street and in the grocery store, as well as at each other's home.

Most people engaged in politics at the ballot box, taking time from work in November and April to cast votes at Meteor School.

Dad and Mom voted regularly, despite the section of bumpy road on the way to our polling place. Back when logging was king, workers laid logs side-by-side over the low, swampy area near Deer Lake. Remnants of the original "corduroy road" still remained. Dad explained that was why the road was so rough, like driving over a washboard.

I grew up with the notion that voting was important and necessary and something people did for their own good. Voting regularly, conversing about affairs, and exchanging ideas became ingrained habits, reinforced by school lessons and observation of the community at election time. After voting, people stayed to chat in the back hall or in the doorway. Men leaned against cars outside to catch up on neighborhood news. Longtime residents Eleanor Evans and Elsie Zesiger often oversaw elections. A small voting booth, or rather a small platform behind a droopy curtain on a rod, sheltered the voter in makeshift privacy. The rough wooden ballot box had likely been used continuously since Town of Meteor's creation.

According to the 1870s survey, Town of Meteor did not exist. Residents on the far edge of Couderay felt their interests were not being represented, so on May 17, 1919, a group of men cast a total of twenty-six votes in favor of separation. They had walked or ridden horseback some twenty miles round-trip to do so. (Less than one month later, Wisconsin became the first state to ratify voting rights for women.)

Town of Meteor successfully separated. Governing the new township became the business of its residents—a duty and responsibility most took seriously. Each thirty-six-square-mile township formed its own local government and elected a six-member town board and chairman. Local government functioned as it was meant to. The people who lived within township boundaries ran the affairs of the community and realized it was an advantage to be involved in decision making. Road supervisors went out after a storm to determine what roads needed to be fixed. Trucks dumped gravel into an impassable mud hole, the grader man smoothed the roads, a snowplow driver cleared drifts of snow, and the elected town clerk kept the records and collected taxes.

Rob, the town assessor for many years, recorded home and farm improvements and taxed residents accordingly. Some railed against paying taxes, but those unhappy with Rob's conclusions had recourse. Anyone who questioned the assessment could show up at a meeting held each spring at the Town Hall's Annual Board of Review. Jerry frequently protested, even though he and Rob were dear friends. Most came with legitimate gripes and had taxes adjusted. Others came, in Dad's words, to "raise hell—just shootin' their mouth off."

Self-supporting farmers made up our community, but some families struggled for a variety of reasons. Collecting commodities and handouts was not looked upon favorably; people expected others to pull their own weight. But neighbors were not without sympathy. When hard times came around for a household, the township assisted by hiring a member of the struggling family for a short-term job, such as putting up or taking down snow fence or cutting brush along roadsides. When men gathered in conversation learned of tragedy, each reacted in silence, eyes shifted to a faraway field; someone reached for a straw to clench more tightly between his teeth; hips shifted, feet shuffled, and caps were adjusted.

A voice offered, "Damn shame, ain't it?"

Heads nodded, eyes cast down, and a second voice extended a long, "Y-u-p." Then they moved on.

Dad took his political turn as an elected member of the Meteor Town Board. He attended regular board meetings; assessed road conditions; met with Cliff Ruch, the town chairman; and helped carry out the functions of the town. About that time, he and others proposed building the Deer Lake Dam to raise the water level and increase the lake's area to enhance fish habitat; the dam would hold back water from Deer Creek. Dad met with philanthropist C. M. Olson of Couderay, who may have provided dollars to help finance some of the project. Before the dam was built, Deer Lake Road led only to the landing. After the lake level rose, Dad and Cliff blazed the path for the new road by using axes to mark the trees. The road still connects existing fire lanes in the Rusk County Forest.

Although men comprised the Meteor town board and the school board, women in our neighborhood also did their part. While Elsie reported and conducted election day voting, she and others gathered together for homemaker's clubs, mothers' PTA clubs, and Ladies Aid. Women led the local 4-H clubs, vacation bible schools, and Sunday schools. Women organized the monthly events at schools and planned neighborhood parties in an all-out effort to create community.

Sometimes Dad and the town board ran into opposition: case in point, the so-called Testy-Tempered Sisters who operated a nearby farm. Whenever the board wanted to widen, brush, or ditch the winding curve on a small section of Y-road, the Testy-Tempered Sisters would have none of it. The road ran near their home and passed by a bucolic shady spot with a public picnic table put in by a local 4-H club. Presumably, the sisters believed repairs were an attempt to take land they viewed as legally theirs. The board believed it was their legal right-of-way, but only maintained the road as they eked out permission, never without a fight. Opposition such as this occurred rarely, or if more often, I didn't know about it.

Operating as a truly local economy for the benefit of its citizens, most people were all in the same boat; the community worked together to create better living circumstances for all. At least that is the way it seemed, seen through the lens of childhood.

Circle of Seasons

THE HAZY DAYS OF SUMMER

For us, as for most other farming families, summer meant Mom canning, hoeing, harvesting the garden, and picking berries and Dad mowing, raking, and stacking loose hay beside the barn in gigantic round stacks. Before balers came into common use, Dad and other men loaded hay onto our wagon using an unusual contraption called a hay-loader, pulled by our tractor. Rotating wire arms gathered up a row of loose-cut hay from the field, lifted it, and dumped it onto the wagon. A pitchfork-wielding worker riding on the wagon arranged the load. Dirt, dust, and scratchy bits of chaff and straw drifted over sweaty work shirts, prickly on shoulders. Now and then, a squirming garter snake got caught up in the hay and tumbled onto the wagon, causing a commotion for the man aboard before he pitched it back to the ground.

Dad let me drive the tractor to help with haying when I was thirteen, but when I was very small, sometimes Dad sat me on his lap to ride with him. Once, he hopped off the tractor to open the field gate, settling me on the seat alone. He left the tractor running, looked up at me, motioned, and jokingly said: "Bring 'er on through." Dad never dreamed I'd been watching his motions as he moved the stick shift. I reached, put the tractor in gear, and it lurched forward. Dad leaped back on in alarm and in time to regain control, no doubt imagining I would tumble beneath the wheels of the moving machine. It would be ten years before he let me drive again.

SUMMER OF THE CYCLONE

On one particular muggy summer evening in 1951, Dad was, as usual, in his after-supper hideaway: the weather-beaten toilet on the hill. The outhouse sat tilting, a left-sided list, with a door that had to be muscled shut, though it seldom was since it stood in a secluded spot. Despite its humble purpose, on a summer's eve it was a peaceful place to relax and survey the leafy domain of home while enjoying the breeze.

On that day, the weather had been cooperative—until the evening waned toward twilight. At home, above the clatter of Mom washing supper dishes and Sam and me rowdily playing, thunder rumbled in the distance. Mom alerted us to the darkening sky. A storm was brewing, blowing in with rapidly increasing ferocity. "I hope Dad gets back to the house soon," Mom fretted, glancing nervously toward the growing fury.

The black sky brightened in bold flashes, outlining poplar trees bowing and swaying madly in suddenly strong winds. Driving rain pounded on the roof and windows. Finally, Dad's rain-drenched shirt could be seen leaning into the wind as he made his way, struggling on the path toward the back door. Wide-eyed, I gasped as rain blew sideways, poured over the back porch in sheets, and drove in under the back door. Dad's dripping figure burst in and water quickly swamped the kitchen floor. Only with Mom's help did he finally force the door shut.

"You kids, get to the basement! Pronto!" Above the howling

storm, Mom's voice sent Sam and me scurrying through the cellar door to the basement below. Mopping the floor was futile, although Mom and Dad tried, and we were relieved when they gave up and joined us in safety.

Beating rain and whistling winds left us wondering what was to come. Then, as quickly as it had begun, the wind and rain receded and left an eerie silence. Emerging from below, we watched through the window as a strange orange hue moved across the sky over fallen trees directly behind our house. "Cyclone," Dad muttered. "Looks like we escaped a bad one this time."

The tornado, what Dad and other locals called a cyclone, had left substantial damage in its windy wake. Dad surveyed the situation, noting downed trees on our hillside. Nothing more could be done at home in the falling darkness. The storm had roared through our hilly farm, and, more troubling, it now swept down the narrow valley eastward, following the usually peaceful tote road along Badger Creek. It was headed straight toward Grandpa Sam and Grandma Bessie's place a mile or so away. Since telephones were still nonexistent in our part of the neighborhood, Dad took up his axe and saw, anticipating fallen trees. He wondered aloud if the road to his parents would be passable at all, intimating his fear of what the state of their place would be. "Don't know what I'll find if I get there," Dad murmured somberly as he left, driving cautiously away.

When he returned several hours later, Dad had both good news and bad. Remarkably, my grandparents and their big frame house remained unharmed. In a twist of fate, the tornado had split in two, parting as it neared, and traveled around their home. Then it came together again and grotesquely twisted, uprooted, and flattened acres of mature, once-towering trees. The destruction was particularly devastating for Grandpa Sam. His proud homestead purchase had included plots of prime oaks and maples—old growth hardwoods. Though he'd traveled all over

Grandma Bessie and Cousin Shirley after the cyclone

the country, the big timber he loved had drawn him to settle here. In addition, Grandpa Sam had meant his trees to be the nest egg of his retirement. Like farmers of his time, he had no Social Security benefits to look forward to. He and Grandma Bessie were safe, but in an instant their future security was at the mercy of fate.

Dad and Jerry wasted no time in planning the attack of a tree harvest far bigger than anything the two had ever attempted, certainly never in summer. The timing, though, was awful. Foliage added tremendous weight to a tree, created poor visibility, and increased danger. Cutting the uprooted tangle of trees presented a safety nightmare. In the leafy camouflage, workers often were unable to gauge the direction of a falling tree, making it difficult to avoid "widow makers"—trees that fell unpredictably. Skidding—horses pulling logs over ground—was nearly impossible without snow and ice. Furthermore, since logging operations normally occurred in winter, finding log haulers and profitable markets was unlikely.

So, why did they do it? No one would allow valuable timber like that to go to waste, and time and weather would soon make it unsaleable. No insurance for the loss of timber existed then. At that time in northern Wisconsin, men still logged much as they had done for decades. They used axes and crosscut and two-man saws, the same types of tools that had been used during the big cutover era. Authentic, hay-burning horsepower also still hauled cut timber. But Dad realized the enormous job could not be done in traditional fashion; he put aside his double-bit axe and purchased his first chain saw.

Compared to today's chain saws, Dad's was a monstrous machine, awkward and impossibly heavy. The steely-blue saw belched exhaust, black smoke, and noxious fumes. Worst of all, it vibrated in literally deafening tones. Ear protection was unheard of then, and the canopy trapped the reverberating sound and exaggerated it until Dad's ringing ears could no longer recover. When his hearing failed a few years later, he attributed the loss to that summer.

Although Dad and Jerry began the timber harvest immediately, they continued only as time allowed. In the fall, when farm work slackened, weather cooled, and leaves fell, Dad hired extra local men somewhat free from their own chores. He also retained a truck, hired a hauler to take logs to the mill, and supplied his team of workhorses as he oversaw the operation, working every day with either horses or men.

Grandpa Sam watched sorrowfully and with great interest as they extracted his prize timber from the tangle, cut it into standard sawmill lengths, loaded it onto logging trucks, and carted it away to market.

Grandpa Sam and Grandma Bessie soldiered on, despite the many difficulties, for the remainder of their lives with their landscape drastically altered—and their nest egg much smaller than they had once imagined.

HUNTING SEASON

In the North Country, no event was more eagerly anticipated than preparation for deer hunting season. Some folks in the neighborhood didn't hunt, but even they were swept up by the excitement and enthusiasm. In anticipation, men put up "buck poles" beside the barn or house—usually in view of the roadside so passersby would be able to admire prizes from the hunt. Since the thermometer dipped below freezing, hunters would hang deer from the poles until butchering. By this time, snow usually covered the ground. If it didn't, talk centered on the possibilities of a perfect snow for opening day—just enough for tracking but not too deep. Dad explained the hunters' wisdom: "If the snow's too deep, deer won't be movin' around much."

And Dad needed deer to be moving both to give him a better chance of feeding his family and to provide respite. A hunkered-down deer population wouldn't do. Hunting season was Dad's vacation, his only break with routine, partial at that. He still had to milk and feed the cows as well as clean the barn. Yet it was the one time of the year when Dad escaped during the day for over a week.

When the hunt began, talk at school turned to who had gotten what. Competition reigned. And even before it began, hunting season turned men to boys again, or so it seemed. Practically giddy, Dad and his partners, Rob, Ole, and Jerry, got together for the ritual of going to buy the hunting licenses. Then, at Rob's place, they made decisions as to where to hunt that year to avoid other hunters and rehashed the laws for the current season. If a question arose as to the exact moment of dawn or sunset—connected to the start and end of legal hunting—or the particulars of a change in hunting regulations, Inee appointed herself their personal librarian. She delved into her encyclopedic collection of knowledge—folded, filed, and stored in her cavernous purse.

Most important, at that pre-hunt gathering, they decided opening morning roles. Who would stand (the men who stood in the woods, waiting for oncoming deer)? Who would drive (the men who walked through the woods, hopefully sending deer running in the direction of the standers)? Where would they meet up midday to make their afternoon plans? Hunters walked for miles, seldom leaving the ground for a tree stand.

Details settled, then came the time for retelling stories and revisiting triumphs—and failures. Their laughter rang out over missing the big buck last year, the hunter whose gun had jammed, and another who had buck fever—freezing in excitement and missing an easy shot as The Big One strolled by. And the prize buck that appeared while "nature called," leaving the hunter only to watch it disappear over the hill. Tales and truths flew fast and free.

Dad and his partners were sportsmen; they prided themselves on taking shots clear of obstruction, never toward buildings or roadways, pulling the trigger only for shots made certain, and trying never to leave a wounded deer without finishing the job. Sometimes that meant tracking a deer for miles. Sometimes they lost the track after nightfall. They had little use for those who poached, illegally hid deer in a brush pile, or otherwise did not follow the rules, or for those trigger-happy men who shot unwisely at the least hint of movement. They frowned on road hunters, those who drove and shot from a car and carried uncased rifles—all illegal acts. They stood in support of game wardens who pinched violators.

That said, in rules unspoken, the neighborhood tolerated "meat hunting"—putting food on the kitchen table by hunting out of season in early fall when deer had been well fed through the summer. Although illegal, this manner of gaining food for a hungry family was accepted practice in the 1940s and 1950s. Few abused the privilege, taking only a young buck, doe, or yearling. Driving past a house with lights on in the basement late at night often meant someone was butchering an illegal deer. Others butchered illegal deer in the barn, where a signal from the house—flicking lights on and off—warned of a nearby game warden. Families who had blabbering children conducted the butchering in secret, after the kids were safely away at school. When I was in first grade, Dad grilled me to assure I could hold my tongue. He and Jerry had taken a deer illegally that fall. If anyone asked questions, my only answer was to be, "I don't know." No one ever asked.

Dad purposefully kept his hunting party small. He preferred longstanding hunters he knew well. Some local gangs of hunters brought in new members with unpredictable hunting habits. For that reason, Dad's party left the immediate area to hunt, preferring to walk miles into the woods around Lost Lake, Bass Lake, or deep in the hills of Deer Lake, all on the edge of the wilderness. Encountering other hunting parties was unlikely. His partners all knew the territory well, relied on their compasses, and could find their way in almost any woods.

Before World War II, no regulations governed the use of colors to be worn for safe hunting. Hunters traditionally wore black and red plaid coats. The variegated pattern and colors provided camouflage in the woods by breaking up the shape of a solid form. When regulations demanded more of the color red, hunters typically pinned pieces of red cloth on the outside of wool hunting coats—or had their wives sew them on.

I knew hunting season neared when Dad gathered his gear on our green davenport in the living room: an extra pair of wool socks, thick felt packs to put inside his boots to keep his feet warm, wool liners for his deer-hide mittens, wool pants, a black and red plaid hunting coat with a pocket in the back for carrying items, a matching wool hat with ear flappers, and a couple of new red bandanas—because you wouldn't want to be waving around a white hanky mimicking the white flag of a running deer's raised tail. Dad considered his trusty wrist compass perhaps the most important item of all, along with a waterproof pack of dry matches to start a fire if lost in the woods and a flare that could be lit to call for help. He laid out his hunting knife, with the sharply honed blade sheaved in its carved leather case, a heavy red box of cartridges, and his gun: a .30-30 Remington deer rifle.

One of the best indications that hunting season was upon us was when Dad bought three giant Hershey bars at Veness's. Sam and I eyed them longingly but knew better than to ask to eat them. Dad carried chocolate for an emergency—quick energy should he become lost. He seldom ate any part of them and shared them with us after the season ended.

A week before season opening, Dad sat at the kitchen table, meticulously took his gun apart, and polished each part with a soft cloth: the wood until it shined, the metal until it gleamed. He wrapped a tiny bit of cloth around a wooden ramrod and cleaned inside the barrel. Then it was nearly time to "sight" the gun at a target. But until then, it remained on the davenport. Sam and I had been instructed to never touch the gun. We knew Dad kept its safety on and chambers unloaded, but we heeded his instructions nonetheless.

When Sam was twelve and Grandpa Sam no longer hunted, Dad gave Sam our grandfather's Savage rifle. He instructed Sam on hunting laws and etiquette, though I believe Sam knew most of it well through those oft-told stories. Steeped in hunting lore by the time I was eight, even I'd heard enough deer hunting wisdom to be a guide. But most girls didn't hunt in the 1950s.

A few days before opening Saturday, the sounds of rifle shots would ring through the air, coming from just beyond the house— Dad sighting the gun! Friday night was filled with excitement. Dad hurried through chores, eager to get to bed earlier than usual. Then long before dawn, lights blazed into the frosty darkness.

Dad trekked to the barn with a spring in his step. Cows turned their heads in utter confusion, reluctantly lifting their bodies from beds of hay. Why were they being milked at this ungodly hour? Dad finished in record time. Mom had been up early too, putting breakfast on the table. Dad wolfed down oatmeal, fried eggs, bacon, and toast, gulped hot coffee, and with few words bolted for the door, gear in tow, off to round up his hunting buddies.

Meanwhile, Mom had been making hunting season plans, too.

MOM'S RETREAT

As she usually did on the first day of hunting season, Mom poured herself another cup of coffee and sat back anticipating the glorious day ahead—without interruption or meal preparation—then plunged into her plans, checking her to-do list. She put another stick of wood on the fire, cleaned up the dishes, and took on the day with glee.

Hunting season always began on a Saturday morning, so Mom sent Sam and me off to go sledding. (Locals hunting on their own properties knew never to fire near a farm site—including our sledding hill.) Mom rolled out the treadle sewing machine to work on outfits for the Christmas program and gifts for the holiday: fuzzy flannel shirts and pajamas. She paged through the Christmas wish book catalogs, looking at toys we'd marked as favorites, and filled out order forms. Taking a break, she stoked the fire and listened to the teakettle simmer on the back of the stove. Mom poured boiling water into Belle's blue teapot to steep the tea, then sat down with a cup to write a quick note to her mother, before commencing again with her secret preparations for the holiday.

By midmorning, Sam and I knew Dad's hunting party had been tramping the woods for hours. Hunters wanted to be settled into place well before dawn; no shot could be fired until then, and no doubt Dad and his buddies had been in place before first light. Throughout the morning, we heard thundering booms fired at unsuspecting deer. With each shot we thought about Dad but knew he was too far away to be heard.

At noon, normally time for dinner, instead of frying potatoes and slabs of beef for another manly meal, Mom ate the meal of her choice: toast and a bowl of soup. She lingered a little longer in the warmth and quiet of the kitchen until Sam and I burst through the door to be fed and warmed. Meanwhile, Dad and company munched on frozen sandwiches, guzzled lukewarm coffee from a thermos, and revamped the plans for the afternoon, to keep moving and try to stay warm. While Dad trekked over snowy terrain and stood in falling snow, his eyes peering for sound or movement, his feet stomping for warmth, Mom reclined for a quick catnap, then sipped hot coffee beside the fire and continued to plan the family holiday.

As the day drew to a close, the three of us wondered if Dad had tagged his deer or if he might be taking his last shot, not a moment after sunset. I pictured deer coming out of hiding, safe for the overnight hours. We waited to hear the sound of Dad's approaching car. If he was late, we knew he could be tracking a wounded deer, dragging a deer from deep in the woods, or the worst thought of all: lost. Dad knew the woods well, so that was not likely. But still? I thought of the tales of Lost Lake and imagined Dad firing his gun three times to identify his location. I silently pondered the prospect of Dad spending the night in the woods—cold, wet, and huddled beside a smoldering campfire. Mom interrupted my dismal thoughts: "Dollars to doughnuts," Mom predicted, certain-sure, "Dad'll have a deer on the car!"

Finally, we heard the sounds of our car crunching snow and we raced to the window. Was a deer draped over the hood? Or hanging from the trunk? Even after dark, the outline of deer antlers could be seen. Hunters kept the rack in full view, especially if it was impressive.

Yes! A deer! With yard lights blazing, Dad and Jerry lifted the deer with block and tackle to a large branch on the big maple tree beside the woodshed. Dad had already field dressed it in the woods, leaving the guts for crows, ravens, and other animals to feed on. He took the heart, liver, and tongue to Mom in the kitchen. She would fry the liver in slabs with onions and pickle the heart and tongue with onions, sugar, spices, and vinegar (Sam's favorite). On this day there was only one buck; no one else had made a kill. Tomorrow all would hunt again, even Dad, who'd used his tag; it was common practice to help each "fill his tag." Dad hurried off to do the chores, milking the cows even this late. Soon he headed to bed for a few hours rest before another pre-dawn awakening, knowing nine days remained.

Perhaps Mom's best days during hunting season came when she sent Sam and me off to school. We knew some of what she'd spent her days doing the minute we opened the back door and were greeted by the sweet-and-spicy fragrance escaping from the kitchen: cherry wink cookies, squares of sugary fudge, fluffy white sea foam, divinity topped with black walnuts, crinkly molasses cookies, and red and white cookies twisted like peppermint canes. Mom had tins of Christmas goodies and packages of fruitcake ready to be sent away in parcels.

The tree by the woodshed usually held a solidly frozen buck, as it did this year, or sometimes a doe, if regulations permitted. In the final ritual of the hunt, Dad, Rob, Ole, and Jerry gathered to take their frozen deer to the registration station. Men stood admiring racks, relaying stories, and recounting hardships as they razzed one another.

After Dad thawed, skinned, butchered, sawed, cut, and wrapped the frozen pieces of venison in waxy white paper, the normal routine began again for both him and Mom. But Mom and Dad's R&R had been a delicious and well-deserved escape. For Sam and me, only waiting for Christmas surpassed the delight of hunting season.

Cherry Winks

Mom discovered the cherry winks recipe in a 1951 magazine. She learned Ruth Derousseau, of Rice Lake, Wisconsin, had been awarded $5,000 as the Junior Winner in Pillsbury's 2nd Grand National Recipe and Baking Contest. Ruth had presented a plate of crunchy cherry winks to Arthur Godfrey (one of Mom's favorite hosts) on his radio show.

1 cup sugar
¾ cup shortening
2 tablespoons milk
1 teaspoon vanilla
2 eggs
2¼ cups all-purpose flour
1 teaspoon baking powder
½ teaspoon baking soda
½ teaspoon salt
1 cup chopped pecans
1 cup chopped dates
⅓ cup chopped maraschino cherries, patted dry with
    paper towels
1½ cups coarsely crushed corn flake cereal
15 maraschino cherries, quartered

In a large bowl, beat the sugar and shortening with an electric mixer on medium speed, scraping the bowl occasionally, until well-blended. Beat in the milk, vanilla, and eggs. On low speed, beat in the flour, baking powder, baking soda, and salt, scraping the bowl occasionally until the dough forms. Stir in the pecans, dates, and ⅓ cup chopped cherries. If necessary, cover with plastic wrap and refrigerate 15 minutes for easier handling.

Preheat oven to 375 degrees. Spray cookie sheets with cooking spray. Drop dough by rounded tablespoons into cereal, coat thoroughly, and shape into balls. Place 2 inches apart on the cookie sheets. Lightly press a maraschino cherry quarter into the top of each ball.

Bake 10 to 15 minutes until light golden brown. Cool 1 minute. Remove from cookie sheets to cooling racks.

https://everyonehasafamilystorytotell.wordpress.com/2016/11/18/family-recipes-cherry-winks

LET IT SNOW, LET IT SNOW, LET IT SNOW

In a northern Wisconsin winter, falling snow perhaps held no fascination for my parents—it meant only more difficulty in doing farm chores. Every Saturday by ten o'clock, when it had warmed up a little, Dad or Sam freed the cows from their stanchions in the barn for their exercise; Mom or Dad turned on the electric switch, pumping our tank full of cold water at the well house; and I accompanied the cows as they took an all-too-slow stroll to satisfy their thirst at the tank. The cows snorted and slurped, blowing clouds of foggy breath into the cold air, their pink noses dripping as they eagerly downed great gulps of icy water. When I drove the cows back to the barn, they rushed to feed at their stanchions where Dad had added clean bedding after Sam had forked the steamy and smelly cow manure outside. With flicking ears, their furry heads excitedly eyed the chopped corn fodder and grain that Dad had scooped into the manger. Once humans had taken care of their needs, each cow's long nose moved up and down, digging in with relish; their jaws ground in a circular pattern, side to side. They chewed contentedly, munching and belching, and then the chomping quieted. One by one, each cow dropped first to one knee, then to the other, clumsily plopping her hindquarters to the floor to settle in on fresh hay until the evening's milking.

Peggy on a sled load of wood

My final Saturday morning chore, the same I had every day after school, was filling the wood box. Deep snow sometimes complicated this matter. I used my sled to haul a load of wood from the woodpile to our porch, but first I had to make a new path. I liked to pile my sled high and take fewer trips. However, that could make the sled too heavy to pull, depending on conditions. Or, an icy trail could dump the entire load. The chunks of hardwood clinked and clunked according to their weight and density as I filled the wood box. Carrying armloads of oak, maple, and ash up the porch steps, I stacked enough to fire both the cook stove and the wood heater until the next day.

When I was finally done, Sam urged, "Come on, Peg! We're wastin' sleddin' time!"

Already warmly dressed for the outside chores, I had sweated through multiple layers. But being damp and sweaty from cow care and woodwork did not matter; now we were free for play the rest of the day!

We lifted sleds from the garage rafters and then took down the skis Dad had made for me. Sam, a stickler for detail, wisely polished his sled runners with fine steel wool until they gleamed. I used a slab of Mom's jelly jar paraffin to wax the homemade skis. Maybe the icy crust under this fresh snowfall would provide perfect gliding conditions, yet keep us from falling through? Alas, no. Skiing was fun, but after falling through one too many times, I gave it up for the rest of the day. Besides, sledding was my favorite.

One year, my Christmas request for a sled was rewarded with a spanking new Silver Streak, a model I considered a Ford, second to the Cadillac of the sledding world, the Flexible Flyer. But the Silver Streak was sparkling. It was new. And it was all mine. I was surprised when Sam informed me, as he assisted me by scraping the bottom of the runners, that the red paint had to be removed. Passing over exposed pieces of gravel would remove any remaining bits of paint and polish the runners to a silvery shine. Gravel served another purpose as well—when speeding too fast down a long hill, the patches slowed us. Dragging a toe to brake was very hard on our rubber overshoes and could also cause the sled to veer sharply off course.

The stiff steering mechanism on a new sled did not allow for the fast turns we sometimes needed. That knowledge was borne home after I felt a strange warm sensation when my errant Silver Streak met a tree and I discovered the warmth was a trickle of blood from a gash to my head. However, once loosened after much hard use, the steering gave maximum prowess for turning quickly on steep icy hills. It was while contemplating this issue, as I rode face down on my sled one Saturday afternoon, that I became distracted and observed the white frost accumulated on the metal mechanism. Why, that frost looked good enough to eat. Lifting my tongue for a taste, "Zow!" I yelped, zooming on down the hill, my tongue stuck to the bar. Forgetting about steering, I ripped my tongue from the metal, leaving a bloody piece of skin still attached. Sledding ended for that day with a *Lesson Never to Be Forgotten.*

On sledding Saturdays, our dog Smokey was as excited as we were, dashing in circles, scooping his nose in snow, and begging to be chased. He raced after our sleds all the way down the hill, grabbing at hats and mitts. When he snatched one, the game was on, the victim tearing across the hillside in hot pursuit of Smokey, plunging through knee-deep snow and capturing the mitten, until it began again.

When Sam or I grew too cold or tired from our sledding and games with Smokey, we burst into the kitchen, wet and hungry. Eager for the warmth of the woodburning range, we shed our wet clothes to dry behind the cook stove. Mom sat with us as we toasted our toes in the radiating heat of the open oven door. We basked, sipping cocoa and nibbling on Mom's ginger snaps, while Mom enjoyed her afternoon tea.

The best sledding Saturday was always after the first big snowfall, when the cocoa tasted sweeter, the kitchen felt warmer, and home seemed more inviting than ever.

WEARING OF THE WOOL

When I was a child, the wearing of wool in winter was standard. The mention of Malone wool, a heavyweight fabric produced from local sheep in a small upstate New York town, would bring

Dad's flash of memory and his proclamation, "Nothin's better 'n Malone pants fer keepin' ya warm in winter winds. Even when yer wet."

People prized Malone pants for their tightly woven, washable weave. Those who wore them said the pants were nearly bullet proof. They kept the winter winds at bay while shaping comfortably to the body's lumps and bulges. Farmers, woodsmen, and hunters valued their water resistance, breathability, and ruggedness, finding them well-suited for work in challenging winter conditions. Dad, Jerry, Grandpa Sam, and Ole wore Malone pants, distinct for thin lines of red and green subtly overlaying thick gray wool. Only Rob differed, choosing deep forest green wool pants and the matching coat, made in Maine for the L. L. Bean label—the same kind worn by game wardens, foresters, and supervisors of the Civilian Conservation Corps.

Wool's facility to hold warmth, even when wet, is attributed to its ability to wick moisture away from the skin, enabling the wearer to stay cozy as a ewe in a downpour or a buck in a blizzard. That's why Dad wore Malone pants all winter, shedding them only when spring temperatures rose precipitously. A worn older pair, with accompanying odors, hung at the door for work in the barn; a newer and fresher pair hung nearby, ready for a trip to Exeland on Saturday nights. Dad's loyal adherence to the wonders of wool was never challenged by the itch some felt when wearing it. Wide and roomy pants allowed for layers of bulky long-johns to shield the skin from any hint of woolly discomfort. Yellow suspenders over the shoulders fastened to sturdy buttons already sewn on, so Mom did not have to do it. Women agreed with their menfolk: Malone pants set the standard for warmth and durability.

Along with Malone pants and some Woolrich clothing, the wooden bins in Veness's were stacked with local wools from nearby Chippewa Woolen Mill in Chippewa Falls. Other wool wear for finer tastes could be ordered from Pendleton in the state of Oregon or Faribault Woolen Mill in Minnesota. But woolen pants were only half of the winter uniform. Plaid wool jackets were as varied as the personalities wearing them. Although some men wore the pattern in green and black, white and black, or blue and black, Dad preferred red and black "buffalo plaid," by far the most common, the type many men wore during deer hunting season. This plaid was so-named after the tartan of Jock McCluskey, originally a Scottish Highlander. McCluskey helped settle Montana, sided with the Sioux at Custer's defeat, hunted buffalo, and became a trader—bartering with Sioux and Cheyenne who traded hides for the rich red of what became known as the large-blocked buffalo plaid, or its smaller version, buffalo check.

The plaid gained universal recognition when the tales of Paul Bunyan made their debut in a 1916 promotional pamphlet for the Red River Lumber Company. Whether in the chill of a northern forest in Wisconsin, Minnesota, Michigan, or Maine, mighty Paul hove his enormous axe to a shoulder clad in buffalo plaid. Ever after, its popularity soared, spreading far and wide like the reverberation from the fall of a towering pine. It's been worn by a long cast of characters: Tom Mix, Roy Rogers, even the Marlboro Man donned the pattern. Elmer Fudd pursued that "wascawy wabbit" wearing an ear flapper trapper hat—of buffalo plaid, what else?

We had a wealth of Scandinavians in our area, few Scotsmen among them, but nearly all were clothed in some version of buffalo plaid. I associated men with the colors and patterns they wore. In Exeland at the local mercantile, a look through the display windows in Veness's identified the wearer. Bright red, green, and yellow Buchanan plaid displayed Jerry's bold personality. Grandpa Sam wore a jacket of subtle blue, its pattern crosshatched in lines of black and gray. My brother wore a soft cream and brown plaid. Only Ole ordered his cheerful, solid red shirt jacks—a nattier cross between shirt and jacket—from L. L. Bean.

When a boy outgrew his nearly indestructible wool jacket,

or parts of a man's jacket finally reduced to tatters, even then the wool's life was not over. Women could be found repairing frayed collars, cuffs, and buttons or attaching round leather patches over worn sleeve elbows. They also purposely shrank wool and sliced it into pieces, downsizing it for another use.

Grandma Bessie created a patchwork crazy quilt with blocks of Grandpa Sam's subtle blue and gray jacket and red and black buffalo check vest; Dad's red, green, and gray Malone pants; and Sam's coat of cream and brown. As I lay burrowed under layers of quilts on frigid winter nights or gathered courage to leap from my cozy bed into the chill of morning, I studied the woven squares of well-worn wool—surrounded by Grandma Bessie's love and the comforting presence of the men in my life.

CHRISTMAS

Prilaman Christmas celebrations rotated among our three families. One year, Sam, Shirley, and I were invited to come to Grandpa Sam and Grandma Bessie's home midafternoon on Christmas Eve. When our parents arrived for the evening, we surprised them by looking like and acting as delivery elves for the gift giving. One of my gifts on Christmas in 1951 was a cardboard canister topped with a metal lid and filled with red, blue, and yellow plastic parts that could be pushed together and taken apart: Krazy Ikes. (Was that the same year Grandma Walhovd wore on her dresses a political pin proclaiming "I LIKE IKE"?). As Grandma Bessie wrote in her diary that day,

> Vera brought down a big roast chicken, cherry pie, fruit cake, oysters, dressing, Jell-O and potatoes. Mr. and Mrs. Walhovd came on the way to Freeman's but stalled on the hill and we kept them here as they were due here anyway. Mrs. Walhovd brought sauce with fruit cake and a loaf of oatmeal bread and with what I had prepared we had a real feast from soup

to nuts. I so appreciated Vera and Freeman's effort to make a good Christmas for a sick aged dad and an old 77-year-old mother. The tears were dangerously near the surface. But I think it was the best Christmas ever. Then Ole came for supper about 6 o'clock. Vera gathered her things and they went home to do the chores. I washed dishes. Sammy and Peggy were so good all day playing with their new toys.

At dusk on another Christmas Eve at my grandparents', Jerry, Millie, Shirley, and Grandma Oberg (Millie's mother, up from her job as a house mother at a college in Milwaukee) gingerly crossed the snow-covered and icy road to join us for festivities. Grandma Oberg busied herself outside for a while, building a pyramidlike structure with snowballs atop a stump near the front door. When darkness fell, she struck a match and lit the candles she'd placed inside her creation. The lamplike glow from her Swedish snowball lantern shimmered through the openings between snowballs, causing my wondering eyes to fly wide open, forever searing the image upon my mind: Grandma Oberg's Scandinavian Christmas magic.

WINTER'S GRIP TURNS SWEET

Through the depths of endless winter, every morning we plunged a dipper through thick, frosty crystals of cream atop our ten-gallon cans, dipping out a pitcher of milk and cream for cereal. Barn cats huddled on the back porch, humped over with tails curled around their paws for a bit of warmth, having left the barn temporarily for Mom's morning handout of hot cooked oatmeal and cow's milk. Smokey attempted to romp, struggling through the deep snow. It took long hours of chores to keep the cattle and horses watered, fed, and warmly bedded in fresh hay. Dad's snowshoes stood just outside the backdoor in readiness, to tromp to the barn or the woods through the next big snowfall.

In the Meteor Hills, winter released its hold reluctantly, as if prying a cold hand from a frozen pump handle. Spring arrived so haltingly in its pace, it was nearly imperceptible at first to any but a veteran of the seasons. Though long resigned to winter, old-timers hoped no one would catch them anxiously peering out for subtle signs of spring. Although banks of snow, higher than car tops, lined gravel country roads long iced-over in packed snow, and morning temperatures still hovered at or below zero, promise was in the air. The once squeaky crunch of snow under heavy boots took on a brittle snapping sound, indicating the sun's rays were lengthening, now warm enough to melt snow that refroze into icy crystals. Chickadees, flitting in frantic, never-ending searches for food, paused long enough to twitter, a sign of warmer days ahead.

For Sam and me, gradually lengthening hours of sunshine literally and figuratively brightened our days, along with raucous calls of ravens, cheerful chickadees, no-matter-what-the-weather-and-always-alarmed squawking blue jays, and flocks of juncos skittering in search of wind-blown seeds. When a lonely red cardinal, scarce in these northern climes, called for a mate, we hoped spring would soon follow.

We noted the woodpile, which had begun in early fall as a mountain of hardwood, split and chopped to fuel our heater, was rapidly shrinking. A five-foot curtain of sword-sharp icicles formed on the south side of our house where heat rose through the ceiling, melting snow from our thinly insulated roof. Dad had little concern for the escaping heat, since he enjoyed gathering firewood; our woods provided a seemingly unlimited supply.

We began to notice that Dad's milking chores took more time now, as cows, which had been dry over the winter, "freshened" when their calves were born; the moms began producing milk again—which also increased the milk check.

Just as people eagerly anticipated deer hunting season in the fall, so they eagerly awaited a thaw, heralding, just perhaps, the arrival of spring. No one happening marked the occasion, as spring crept in with baby steps, then disappeared, almost unnoticed. But when sap dripped from recently broken twigs and the drips formed miniature icicles by morning, it signaled spring was surely on its way.

Sap also meant local farmers turned their eyes to their woods for another source of income. Before I was born, Grandpa Sam made maple syrup in a metal pan over a fire deep in his woods. I heard stories as we walked to his old sugar bush—the maple grove—to view the flat, red rocks gathered around a fire pit. Our neighbors continued the tradition. I was six or seven when the lure of Rob and Inee's sugar camp won me over completely. For a few weeks during each spring's sap run, they lived and slept in a large, walled army surplus tent in the woods, preparing meals over an open fire. Rob set up camp near groves of maple trees. Inee set up living quarters, adding a small table, cots, and cooking supplies.

Using the hollow core of soft-centered sumac wood, Rob whittled his own spiles or spigots—the tiny tubes that drained sap from maples after the trees are tapped. To tap the tree, Rob drilled a hole and inserted his spile. The holes did no harm to the trees, since each year they closed over the course of the growing season. Sap drained into buckets, then Rob and Inee boiled down the collected sap until it thickened into amber-colored syrup.

Dad found sap tasty, like water with a faint sweetness. Like many, he drank it right from the tree or pail, proclaiming it an excellent spring tonic. Some claimed a healthy glass or two of sap good for the constitution. I was amused by the stories but preferred the cautious approach, never tempted to drink more than a small sample after having heard tales of older schoolboys' pranks enticing an unsuspecting victim to drink too much. Sap made an excellent natural laxative, both fast-acting and effective.

We were invited to watch Rob and Inee's syrup-making

process one spring Sunday. The camp fueled my ideas of survival and the romantic notion of living in the wilderness. After visiting, I became enthralled with pioneer stories. Sugar camp was the ultimate pioneer girl's dream, so when our family joined the ranks of maple syrup producers, I was thrilled. At first.

Dad's decision to go into the syrup-making business began in the mid-1950s after Albert Halberg, who lived across the road from our mailbox, met success with his sugar bush. Our farms were small; an additional cash crop to supplement the milk check was always welcome. Although maple syrup production was labor intensive, like farming and logging, it came at a less demanding time of year. Dad happily would put his woods to yet another use.

My enthusiasm faded when Dad brought home five hundred second-hand buckets he'd bought at a bargain rate. The pails had been used for cherry picking, holding up to thirty pounds of cherries; Dad thought they'd work well for collecting sap. He only had to drill a hole in each shiny bucket to hang it on a tree. But first, the buckets had to be washed of gummy cherry juice. That job was up to Sam and me.

In the middle of winter, the job had to be done inside. Our farm kitchen was no showroom model, but it had never held five hundred awkward pails and lots of water handled by two crabby kids. Instead of sledding, we spent the weekend cleaning. Dad got the operation underway. When he was through with his instructions, or he could no longer stand our grumbling, he returned to other chores, leaving Mom to supervise. We were careful not to complain too loudly, since the mantra of the day was "kwitzyer-bellyachin'"—meaning all of us were expected to swallow hard and do our part. Eventually, we finished the disgusting job. The sweet—now sickening—cherry smell still remained. The only thing keeping us from sticking to the floor was the water sloshing about. We did our best to clean up, still feeling ornery and with little joy of accomplishment, too sullen to be glad we were done.

Little did we know that this messy job was but a foreshadowing of the messy work that would happen every subsequent spring.

THE RUN

The calendar heralded March 15, and the woods remained waist-deep in snow, leaving only one way to navigate winter's accumulation. Strapping on his longest set of snowshoes, Dad took to the woods in earnest to tap five hundred maple trees. Although Rob had whittled his spiles from sumac wood, Dad took a modern approach, purchasing metal ones. He carried them in an old bucket, along with a hand-operated brace and bit. Dad hoped the sap, which was still frozen solid, would "run" soon, but there was no way of knowing when a warm-up could occur and a sudden run might begin.

After tapping, Dad worked on getting the syrup operation ready to take on the *big* run we knew was bound to come. He'd already built the wooden evaporating—or boiling—shed, a long

Boiling sap in the shed

building with a high-pitched roof. Dad and Sam stacked its open end high with five-foot lengths of hardwood, ready to feed an immense fire. A gablelike upper box on the roof had sides that could be opened by block and tackle, to release steam rising from the building. Dad would store collected sap in tanks on the hill just above the shed; gravity would drain the sap through a movable pipe into a series of open boiling pans at the back of the evaporator. Dad had purchased this metal commercial evaporator from Anderson's Maple Syrup in Cumberland. It was perhaps twenty feet long and four feet wide, an open pan with deep flues extending down into a roaring firebox below. Cold sap heated to a rolling boil, evaporating as it flowed through the boiling pans, from the back to the front finishing pan. At first, the sap looked like boiling water. As evaporation continued, it thickened and bubbled more slowly.

The entire process had to be carefully tended. Running out of sap would ruin the evaporator; likewise would an uncontrollably hot fire, by turning syrup into a mass of solid maple sugar. Syrup couldn't be runny, too thick, or cooked too long and taste burnt; and it must never boil over. A hydrometer measured the syrup's density, ensuring proper thickness; to keep it from foaming too much and boiling over, the person manning the fire added a drop of butter.

To bring sap from the woods and navigate ever-changing weather conditions, Dad would first mount a covered tank on top of a dray—made of log skids—that rode on top of crust or deep snow. Later, he'd use a wagon. Queenie and Joe pulled both the dray and the wagon. Our horses had spent the long winter months lounging in the barn eating hay and would have a rude awakening when suddenly pulled into service. Like Dad and the workers he planned to hire to help haul sap from the woods to the storage tanks, our horses would suffer sore and strained muscles through the exertion of slogging through snow, slush, and then mud.

Moving about the woods to collect sap in the extreme weather of early spring would be a constantly shifting challenge. Piles of snow one day, deep and bitter cold the next, with freezing, driving rain another day. Then, in a sudden turnabout, a spring thaw.

At first, workers could balance atop the snow wearing snowshoes while carrying five-gallon pails of sap in each hand to the dray. But as snow melted and settled, it became soft, unsuitable for snowshoes, and no longer able to support a man's weight. When freezing and thawing occurred, a thick crust formed on top of the snow. Falling though the crust with every step was exhausting. As the season wore on, eventually snow and frost left the ground and the old logging roads in the woods filled with deep slush, standing water, then mud. Wool clothing and footwear would become soaked.

No matter the weather, it took forty gallons of sap to boil down to one gallon of syrup. It required one trip through the entire five-hundred-bucket sugar bush to collect enough sap to begin the evaporating process.

People preferred sunny, clear, dry weather for boiling, since humid days slowed evaporation. One promising morning dawned crisp and sunny with a balmy breeze. On my way to school, I surveyed Halberg's evaporating shed across the pasture. It stood in readiness, but no steam escaped from its roof yet. A large sugar maple served as the corner fence post of Halberg's pasture, barbed wire circling its girth. Covered sap buckets hung on the tree's two spiles. I noticed a tiny stream of sap moving slowly down the spout, forming a drip at the end, then dropping, *kerplunk*, into the empty bucket as another drip slowly formed.

When I returned from school that afternoon, I heard the steady *plunk-plunk* of sap dripping into a rapidly filling pail. I hurried to peer into it—it was about to overflow. Halberg's shed now showed signs of activity and curls of smoke rose from the chimney, signaling the run was underway!

At home, sap collection was in high gear. Dad had spent the day hauling and began boiling that evening. We watched our production for the first time: cold sap warmed quickly to a rolling boil, and moist steam filled the room and rose. Dad opened the gable sides of the vent, releasing billowing clouds of steam into our valley, turning the air sticky-maple sweet.

Sam, Mom, and I milked the cows and did the other chores, checking in on Dad's work in the shed. Dad boiled late into the night, as he had enough sap to stay ahead of the evaporator. We whooped with joy as the first sweet gallons of high-quality amber syrup drained from the finishing pan, filtered through thick felt to remove impurities, and slowly dripped into ten-gallon milk cans.

Then we hoped for cooler weather to slow the season down. It did. Temperatures dropped, and snowy ruts in the road froze hard overnight. But the day after promised to be above freezing. Sure enough, next morning Halberg's pails thawed quickly, sappy drops already dripping into the bucket. That afternoon, a faint sweet smell drifted through the air. Halberg's shed steamed at full blast. I hurried home to find Dad had hired a neighbor to man the evaporator fire during the day while others gathered the rush. Pails on the trees were running over, losing the rich sap before it could be brought in. Dad would boil all night to keep ahead of the run.

Though I was too young to help with the hauling or boiling, I dragged the green metal lawn chair into the shed beside the fire for Dad, never realizing that sitting down to rest was unlikely. He remained through the night, feeding the firebox monster with huge chunks of wood. The fire growled as it consumed the woodpile bite by hungry bite.

Although I've heard yarns spun of starry-eyed folks pouring freshly made steaming syrup over just-fallen snow, this quaint activity did not happen at our place. Possibly because after smelling syrup cooking for days at a time, it was the last thing we considered a treat, no matter *what* we poured it on. Though we did make maple sugar candy occasionally, we did it without fanfare, and no one was crazy about it.

Through hard work, uncertainty, unpredictable weather, lack of sleep, and equally strained tempers and muscles, Mom soothed everything, helping out with chores as needed. She provided meals on the run, watched the firebox, and made sure the daily farm chores were taken care of. Sometimes visitors and relatives overran us. I set up log benches in anticipation of guests who came to watch the proceedings. It was fun to entertain curious friends and folks from town who wanted to see what the syrup business was all about.

Gradually, as snow melted and hard crusts softened, Dad announced he had seen tiny snow fleas moving about on the snow. Those wriggling springtails had hatched in late winter from their home in the bark of a dead tree. Soon enough, sap no longer migrated to the roots of the tree each night; the tree needed it high up for the unfolding of tiny buds into fresh, green leaves. As a result of warmer weather, the season trickled to an end.

Syrup seasons varied from short and unprofitable to long, profitable, and exhausting. When the season was over, every member of our weary family heaved a huge sigh of relief. We packed pails away cheerlessly, scrubbed the evaporator, and covered it until next year. Tired horses, on a rare day of rest, lolled lazily in the sunny pasture, eyes toward the barn, hoping Dad would not come to put them into harness.

Then came the business end of the business. Because of impurities in late season sap, the syrup became dark and stronger flavored. Dad sold this syrup in five-gallon bulk containers directly to larger producers. It supplied one ingredient in concoctions sold as maple *flavored* syrup. Due to its darker color, companies often sold it in tins or brown glass. Dad sold some of our high-quality syrup to local residents for the ridiculously low rate of five dollars a gallon. We used clear glass bottles so the buyer could see the fine amber color.

Syrup season seemed in the distant past when one day in midsummer Dad announced his plans: we would bottle some of the best syrup into smaller-than-gallon sizes. He would pack the smaller bottles into cardboard cartons and deliver them. That meant bottling the syrup in our kitchen during the already-sticky summer days of July. Sam and I had to help (of course). We soon learned this task made washing sticky cherry buckets seem like a Sunday school picnic.

Small stores in the area and Ma and Pa summer resorts bought our bottles of syrup by the case and then sold it to locals or to tourists. Each bore a label in green print. Dad added *100%* because people often asked, "But is it *pure* maple syrup?"

100% PURE

MAPLE SYRUP

FREEMAN PRILAMAN

Exeland, Wis.

And I would have added:

ASSISTED BY HIS HARD-WORKING FAMILY

(AND QUEENIE AND JOE)

We heaved a gigantic sigh of relief when we filled the last bottle, then forgot about it through fall and winter until the next spring when. . . .

Over an extra cup of his strong lumberjack morning coffee, with plenty of cream and sugar, Dad announced it was time to take the buckets from storage. We greeted the news with mixed feelings of excitement and dread. It meant, yet again, unpredictable long hours, backbreaking work, and stress, all added to the usual farm labors. For us kids, however, there was also an air of anticipation: the promise of extra cash, escape from the routine of farm life, visitors, and an acknowledgment that, yes, another spring was on the way.

FINALLY, SPRING!

When we saw the first sign of spring thaw, we turned our attention to one of our favorite activities. Outfitted in tall rubber boots, preferably without leaks—or at least well-patched—and armed with an ordinary stick, we could spend hours in the pursuit of "ditching." We could ditch anywhere. A country road would do, but the best place was at the base of steeply surrounding hills in a watershed. Our dirt driveway in the valley beside Badger Creek was perfect, running with water in every direction. With long sticks in hand we ushered the melting slush and water from what had once been frozen snow piles into watery paths flowing to still-silent Badger Creek. We shouted warnings to one another as floating pieces of ice, snow, and slush broke loose and headed down our temporary rivers: "Break out! Flood downstream!"

It wouldn't be long before the deeply drifted bed of snow filling Badger Creek to the top of its banks began to soften and sink lower. Ever so slowly, a dark opening appeared somewhere in the blanket of white. Under the nearly transparent layer of ice, a bit of running water appeared, as if dancing playfully in mysterious patterns under glass. Then, on a rare spring day we heard it . . . the first faint trickle of flowing water! When the snow-filled creek began to discolor and turn slightly yellow with slush, we knew the creek's opening was imminent. A heavy rain could make it happen suddenly during the night, awakening us with its surprising roar as water hammered over boulders in a noisy rush. But most often, Badger Creek's thaw came in short stretches. We eagerly watched bits of ice and snow break away, changing a little day by day.

One April afternoon, Badger Creek put on a stunning once-in-a-lifetime performance. On an unusually warm day, I was ditching near the creek. Though we had heard the faint sounds of moving water, deep snow still locked the creek, bank to bank. While creekside, Mom, Sam, and I heard a tremendous roar upstream and saw flashes of a churning, yellow-brownish wall of water rushing downstream. We retreated to the steps of our back porch to watch the torrent fill the creek and rise to its very banks. Tumbling chunks of snow and ice and slush were swept away in an instant. For a moment we thought the creek would overflow, so we prepared to scramble to the highest step. But mesmerized by Badger, we remained frozen in place and watched in astonishment. In seconds, cleared of winter's frozen matter, the creek subsided. Within minutes, Badger Creek had completely opened and was running at its normally high spring flow.

"Why, I remember the time Badger Creek flooded," Mom suddenly remarked, "after a big spring rain in 1940. We'd just picked up our baby chicks from the depot, not more than a week or two old. The chicken coop was over there, just across the creek." Mom pointed to where I had often played in the summer, then continued, "It had been raining all day, the creek raging. We thought it would go down in the night . . . but it rained all night long, too. In the morning, the creek surrounded the chicken coop and swept away our baby chicks . . . one by one." Mom paused, lost in the image of her chicks bobbing downstream. "We couldn't do a thing about it . . . there was no money to buy more . . . we went without Sunday chicken dinners for company that year . . . and didn't can any chicken for the winter either."

Such was the unpredictability of Badger Creek. Though it flooded often, we usually thought of Badger as our friend, simply providing clear spring water to drink—and entertainment.

In another unusual year, while several feet of snow still covered the woods, the temperature soared to eighty degrees. Water ran with abandon, filling and overflowing ditches at the side of the road and flooding across gravel roads. Company had come to visit; all were enjoying that magical spring day. Even the grownups played in the water. Dad, armed with a shovel, cautiously eyed our guest who stood poking at snow chunks overhanging the fast running stream. Many years of experience and the twinkle in Dad's eye told me that he knew the man stood on the precipice of disaster. Sure enough, the snow gave way and he tumbled into the drink. With peals of laughter all around, including from the victim, everyone became a kid again.

In contrast to the quiet of the winter season, we now listened to the roaring creek. Our spirits soared with consistently warming temperatures, and rain fell, as gentle as the nighttime moths that flew about in abundance. The air took on the rich smell of earth from woods carpeted in a matted, leafy, brown bed, with tiny bits of green poking through.

Yet, just as the snow and ice of winter hampered travel, so did the mud and ruts of springtime. One day, Ole walked over for an evening visit to announce that weight limits had been placed on the roads. That meant heavy truck travel was limited because it made the roads worse. Country dwellers, having had mud-driving experience, could navigate the worsening roads for a while longer. But sooner or later, one day Dad would proclaim, "Better stock up and plan on not goin' to town for a while, Ma. Road's breakin' up." Then trips to Exeland, Birchwood, or anywhere else were out of the question.

Folks who came to visit might drive their car as far as conditions allowed, then walk the rest of the way. But inevitably, sometime during the breakup, people who lived along Highway C would be called upon for assistance. Then Albert, Fred, or Ole would report with a hint of annoyance, "Hadta get outta bed after midnight and getta tractor to pull sum guy outta t'mud hole t'other night. Sum damn fool from town; buried t'car inta mud

right upta t'axles." Then Albert, Fred, or Ole—and his buddies—
chuckled in amusement, with the slightest air of superiority that
only another farmer would understand. The rescue effort had
been well worth the time for its entertainment value alone.

Some years were worse than others; as many as three or four
weeks sometimes went by before the roads finally dried up and
became passable. But surely and ever so slowly, mud dried up as
spring inched in.

Then one evening from the pond north of our house came an
ear-splitting trill: the joyous and welcome song of spring peepers.
Frogs. Spring was *here*! (Could slithery tadpoles be too far behind?) The arrival of the frogs would be quickly followed by other
jubilant proclamations of spring: walleyes running and suckers
spawning; fluffy new kittens hiding in the barn; schoolboys' excited voices resounding on the playground, "Holy cats! Time ta
go trout fishin'!" And Mom's flush of spring-cleaning springing
into full force.

One glorious day, Mom would throw the windows and doors
wide open to clear out stale and smoky air and soot. Mom's paint
brush appeared, touching up the kitchen with a new coat of sunny
yellow. Freshly washed gingham curtains danced on the line.
"Get outside and air out your britches," she'd happily command,
quickly followed by, "Don't slam the screen door!" and then, "And
remember to clean your feet; don't be trackin' mud on my just-
scrubbed floor!"

Winter-bound calves and heifers were as happy as I was to
be outside when green grass first appeared; they cavorted in joy,
twisting and turning in pure ecstasy when released onto new
pasture. Cows contentedly grazed on shoots of green, sunshine
or not, their grimy winter coats washed shiny clean in gushing
spring rains.

When I went exploring outdoors again, although too early

for mosquitoes, a new crop of blood-sucking wood ticks awaited me atop every twig or blade of grass. We all remained continually on the lookout for them, picking them off our cats and Smokey, but they didn't stop me on my woodland wildflower walks. In high school, when I read *Walden*, I remembered that as a child I had appointed myself Commissioner of Wildflowers. As Henry David Thoreau himself once suggested, I combed the woods every day in May to search for the first welcome mayflowers (hepatica), followed by bloodroot, Dutchman's britches, spring beauties, marsh marigolds (cowslips, to us), pale purple violets, nodding yellow trout lilies, and the rarer white ones that I knew grew only by the pasture creek.

On one of these wildflower walks, in pursuit of the wood anemone and fragile maidenhair fern I knew grew in the coolness of the spray of splashing water, I clambered up the creek bank to the rose-colored rock wall where it rose to a height of twenty feet above Badger Creek. As expected, the plants clung by thin tendrils there, finding purchase in the red wall's thin-layered cracks. And, unexpectedly, there I spied it. Tumbling in the white frothy water below, where Badger Creek was forced to make an abrupt turn at the unyielding wall—one of my errant baseballs, sent downstream last summer after it fell into the creek? No! A large, perfectly round rock, spinning, spinning, spinning. How long had it been caught in that rocky corner, trapped by the current, swirling on the solid rock of the creek bed? And where had it come from? I took the mysterious rock home to add to my collection: a treasure even surpassing that day's wildflower finds—which had also included Dutchman's pipe (wild ginger) cozily hidden under the guise of brown leaves. Even Dad didn't know about that plant's location until I dragged him and Mom to see. Neither did they know anything about the rock. All in all, a satisfactory woodland walk.

In late May, when snowy white trilliums triumphantly carpeted our woods and replaced the melted snow, I knew that spring had taken hold at last.

Mom knew, too. Early sprouts of tomato and green pepper plants rose from her wooden flats of garden soil in our sunny east window. Then a brown envelope arrived in the mail—Mom's Burpee seed order. She retrieved the sealed jar of left-over seeds from the cool basement, placing them alongside the new packet containing dill, carrot, radish, and lettuce seeds. Over the kitchen table, Mom wondered when our heavy clay soil would be dry enough to plow. Meantime, she sorted the best potatoes and cut eyes for planting and bought onion sets and pea seeds in Exeland. She could plant all of those plants in cool spring soil. Beans, squash, and pumpkins needed later warmth.

With ditching season now over, Shirley and I dreamed of just one thing—mud pie–making season. Badger Creek, on both Shirley's stretch and mine, trickled by with an ever-ready supply of water, sand, and mud, offering various consistencies for our favorite warm-weather pastime. But we would have to bide our time until Badger's spring flow subsided and finally allowed us to cross the creek on piles of sand and step-on rocks. Scratches and bruises would not be far off.

Mothers stocked up on home remedies for cuts, scrapes, bumps, and the plethora of mosquito bites sure to come. Before long, the parents and kids of farm families would ease into summer chores, working in the garden, getting the cows, and making hay.

But for now, spring had finally come!

Escapes

NATURE'S PLAYGROUND

Oh, the niches of childhood. Those magical carefree days of sunshine-filled dreams and escape, anything-is-possible innocence, games, and rock-hopping. Even a boulder midstream could become a huge and elegant castle in my mind. An oversized cardboard box could morph into a hermit cabin in the woods. This retreat, originally the packaging for our first real refrigerator, lived out its usefulness in a week or two. It crumbled away in the first hard rainstorm, as fleeting as my imagination. No matter—I was ready to move on to other imaginary lands, anyway.

Imagination could make me anyone: bold as King Arthur, regal as the Knights of the Round Table, boisterous as Little John and Robin Hood, and tiny as the Teenie Weenies. I could escape to tropical summer climes or burrow into my ice cave, in real or imagined raging Arctic blizzards. My bedroom on the north side of the house was far from the evening's glowing wood fire. But as icicles hung long and glistening outside my window overlooking paths trodden in deep and drifting snow, I remained warm and snug. As I lifted Grandma Bessie's heavy patchwork quilt above my head, my bed became a tent big enough to include my companion, a striped tomcat named Tabby. The two of us weathered many a dark howling night, cozy together. Despite the baying wolves, frigid blasts, swirling snow, and subzero adventures by dog sled, we remained huddled in comfort, awakening only to an early morning winter breakfast of sizzling bacon.

When the seasons changed, Tabby and I climbed aboard the raft with Huck Finn and Jim, afloat on the Mississippi. A lighter bedcover sheltered us from summer's sweltering sun. Sometimes we ran aground or pulled up on a sandbar or an island for a bit of fishing and adventure. We played hooky with Tom Sawyer, climbed through the window with Huck at the Widow Douglas's house, and marveled at the astonishing gall of the Duke and the Dauphin.

Indoors or out, my imagination ran as wild as I did. In summer, I buckled on my brown leather sandals with the small cutout patterns. They protected my toes from snakes, stones, and stubbed toes and could be quickly shed for a dip in the creek or going barefoot. At $1.98, Mom could afford the numerous pairs I blew through each summer. Slamming the back-porch screen door, I would make my escape.

Rock hopping across Badger Creek, I entered a fantasy world inhabited by Benny, Jessie, Violet, and Henry of the Boxcar Children. In the wooded area, I could easily create the Boxcar World—using chunks of oak from the nearby woodshed for stools and table, and green pine needles on the forest floor for cots. An endless supply of items salvaged from our dump nearby made my boxcar home complete: tin cans, old coffee pots, a cup with a broken handle, and a coffee-can-lid plate. Each time I had a hankering for change in my arrangement, I redecorated by adding new "finds."

Summertime also became Mud Pie Time. Badger Creek, the dump, and an occasional new, metal tea set provided everything necessary. Pebbles from the creek, soil, sand, mud, and

Peggy's mudpies

water—mixed to measure—filled my brightly painted tin pans with cakes, cookies, pies, and other delectable bakery goods. Time flew by as I concocted new shapes and variations and scraped failures from pans, swishing them from my counter—a weathered plank next to the stream. I reformulated my batter until each product was sun-dried crisp and golden brown. Once baked in the afternoon sun, they remained on shelves until I sold them to imaginary customers. If a night rain had cleared my shelves, I just began again.

There were so many natural play areas. Every afternoon, the meandering Badger Creek became a flash flood or raging river.

Sam's cowboy hat and Peggy's new boots

I forded it in my covered wagon, and only *I* knew that our deep ravine was actually a hidden and remote box canyon, inhabited by dangerous horse thieves who foolishly believed no one knew they were there. Luckily, my rapid-fire six-shooters, stashed in Annie Oakley leather holsters tied to each leg, fired caps fast enough to save me from those thieves (my fearsome black and white cowboy boots helped, too.)

Mom and Dad had considered cowboy boots a luxury; on this we differed. One year, I picked green and yellow string beans for three or four cents a pound, depending on size and grade. Although weary of bending to pick bucket after bucket, carrying them to the end of the long row, and dumping them into long red mesh bags, I kept my eyes on the prize. The Stokely-Van Camp company representative weighed my bags, loaded them on a truck for canning, and paid me. By the end of the four long days, I'd picked over two hundred pounds and earned enough for my dream boots—$4.99. I clunked around in them daily on my way to get the cows from the pasture for evening milking. Good thing I did.

Herds of cattle stampeded regularly—oddly enough, every afternoon about four o'clock, not long before milking time. Matt Dillon and his sidekick, Chester (limping along on his bum leg), came along to save the day (Miss Kitty at the Long Branch Saloon had told them there was trouble out on the farm in Ravine Valley!). They were pleasantly surprised to learn I had handily taken control. Doc Holliday often drove up in his horse and carriage by the time I'd driven the cows into the barnyard. I didn't always need his services, but sometimes he would transport a bad guy back to town—someone *I* had already securely captured and hog-tied, of course.

My escapes could be anywhere: on top of a rock pile as Dad cultivated corn nearby, under a giant maple tree away from summer sun, atop the seat of a scrapped piece of machinery, on the old wooden-wheeled wagon discarded in the pasture, and even in the manger of our cow barn. Amid the sweet, musty smell of hay, cows peacefully chewing their cud, cow stomachs growling, and cow mouths belching steamy breath, I sank into corn fodder and dreamed up adventures.

Time moved incredibly slowly. While I eagerly waited for the weekly Saturday nights in Exeland, bimonthly Sundays at Grandma and Grandpa Walhovd's house in Birchwood, or my own birthday, time took its time. Finding escapes? Essential. What else was a kid to do?

JUNKED

When metal was in demand and the price of scrap iron high, Harry Taylor, the local junk man and jack-of-all-trades, went on his rounds. Farmers happily watched Harry's truck hoist and haul away their tanklike unusable cars, earning both Harry and the farmer a few cents a pound. But often, the farmer junked his car himself.

Every farm had an abandoned car on the back forty or hidden behind the barn. Bumping through the lumpy pasture to a resting place, farmers left their forlorn cars sinking into the ground. Ancient rubber tires stiffened and disintegrated into the soil on brittle rusty rims; wooden steering-wheel parts crumbled; wood and metal tire hubs lay in fragments on the earth. Solid and hulking, battered by wind and time, only the faded reddish-rust of weathered iron and steel remained. Some cars sat atop rock piles or had saplings wedged through sagging bumpers and growing toward sunlight at the edge of the woods—leaving kids to wonder how Grandpa ever parked them there in the first place.

Some cars were as square as a cracker box; some had sweeping back seats and high fenders; others, such as Grandpa Sam's Durant Star Car, were open-air affairs—the body built as if atop a wagon frame. Grandpa Sam hadn't driven that car since Jerry and Freeman had bought cars as young men. Years ago, Grandpa Sam had driven the 1920s model to an opening along the tote road and left it there, where it remained moldering into the ground near the garden path. While our grandparents worked in the garden, Shirley, Sam, and I entertained ourselves for hours, climbing over the huge rotting tires and taking turns in the open seat at the wheel, pretending to crank the old Star to life.

One of my favorite abandoned cars stood gallantly at the Halberg's place: an old Nash with an enormous snout towering above our heads. I spent many happy hours playing in that dilapidated car, along with my friend Richard and his little sister who tagged along. (Carol still insists we ordered her to stand behind us in the back seat watching as we drove, but never, *not even once,* allowed her take the wheel. Sure, the Nash was in drivable condition—but not for a four-year-old.)

On hot summer days, we drove leisurely, cooling ourselves by hand-cranking the windows up and down and adjusting the side window vents, moving them side-to-side for enhanced airflow. Though it didn't work, we flipped the front windshield's tiny cooling fan switch—on, off . . . on, off—imagining breezes lifting our hair to the wind as we cruised swiftly away. Of course, even the fresh air did not take away the car's peculiar smell; a musty odor rose from plush velvet-upholstered seats. Moist or not, we bounced gleefully on them, both front and back, until the thick-coiled springs threatened to break through the fuzzy brown covering.

Waiting for my turn in the driver's seat, I watched Richard joyously spin the plastic Bakelite knob on the steering wheel, precariously maneuvering the car around make-believe corners. Screeching on two wheels, we sped through cows grazing in the pasture and pictured the bewildered herd scattering in every direction. In truth, the cows stared at us in dumb curiosity, as if we'd completely lost our minds.

No Sunday trek to the Mulkey place was complete without a stop at the ancient jalopy left alongside the hayfield. It sat with sagging springs and a rusty-brown folded hood flung open, as if waiting for the mechanic who never arrived. The open-topped touring car was too far gone to sit inside and play; I could only imagine speeding away in that once-classy machine. Overtaken by climbing grape vines, young maple trees, and weeds creeping through the floorboards, the junked car sat moored forever in place. Leaving it behind and hurrying on, Sam and I hoped Mom would have enough time to make a rhubarb pie for supper, a fitting ending for our day away.

Imaginations stoked, we riders entertained ourselves by stealing nervous glances in the rearview mirror, watching for the red lights of cops in hot pursuit and yelling out, "Faster! Faster!" Or warning, "Bad corner ahead! Grab the ceiling strap above and hang on!" (No seatbelts here). "Careful of bouncing too hard—that solid wood-paneled dash is not padded." Grumping about the driver's abilities while hoping he or she would finally relinquish control, we amused ourselves with gadgets: adjusting radio buttons, mirrors, and antennas; opening and slamming glove boxes; turning on immovable windshield wiper blades.

Finally, we lost interest in those car rides; maybe the field mice moved in, the seats rotted through, or Harry Taylor carted away the Nash for scrap. Or maybe we just grew up—traveling on down the jostling road of life toward adulthood.

A BUTTON BOX DAY

One particular rainy day offered a very different kind of escape. School had been out for a couple months. I had no more books to read, and I'd already read the parts that interested me in Mom's magazines and even Dad's hunting magazines, so I had to wait for new material coming in the mail.

After counting my piggy-bank savings, folding my doll clothes and tucking them into a box, and rearranging our game cupboard of Tinkertoys, Tiddlywinks, Pick-up Sticks, Dominoes, a bag of Jacks, and my Cootie game, I was at a loss for what to do next.

The storm had begun in the night, sharp lightning brightening the sky, crashing thunder rolling from hill to hill and echoing down the valley, followed by loudly drenching rain. By early dawn the creek roared. Dad had donned the army surplus rubber rain gear he wore to go fishing whenever he expected the sky would likely open up with torrential rains. This morning, he'd worn it to find the cows sheltered in the far corners of the pasture, bringing them to the barn while rain fell in sheets. Now left on the porch to dry in muggy air, the raingear still hung heavy, limp, and wet.

I'd listened to the sounds of Badger Creek rising in the night, water tumbling noisily over rocks and boulders. Water covered the wide stones I normally used to cross over to the other side. Water in the creek had risen so high it would be several days before I could use the stones to leap across to my mud pie–making area. White, frothy water foamed higher in pools where water pounded over a fallen log. Brown water poured from the pond north of the house into the main stream. Puddles on the driveway overflowed, discharging muddy rivulets into the rising brook, and still the rain showed no signs of letting up.

"Well," Mom suggested, "how about sorting the button box?"

This was an age-old entertainment—a standard escape at Grandma Bessie's house, too. Every house had a button box. Whether in an actual box, a large glass jar, or a decorative storage tin, they were all "button boxes." A collection of buttons of every size, color, shape, and material gleamed through the fat glass jar that was ours.

"How about finding some green buttons for my new housedress? And while you're at it, you can find some big, black four-hole buttons to sew on Dad's new overalls, so he can fasten his suspenders to hold up his britches," Mom quipped.

It was a deal. I sometimes added to our supply whenever Mom turned clothes into rags. It was my job to snip the buttons from Dad's holey work shirts and Mom's worn-out housedresses or tattered blouses. Every house had a ragbag, too. Saving buttons, zippers, and bits and bobs of cloth for the next sewing project was part of what we called "reusing." Dad's long underwear found its way into the mop head, clamped and held in place. Strips of flannel from my old pajamas, sporting pink elephants, now washed cows' udders. Used buttons showed up on future aprons, housedresses, or shirts. Today they would provide rainy day entertainment.

After we ate our dinner at noon and cleaned up the dishes, Mom sat down to read a magazine. I dumped a mound from the jar onto our oilcloth-covered kitchen table, and the sorting began. Buttons large and small, round and square; two-hole, three-hole, and four-hole; fancy and plain; white, black, red, yellow, and blue. Buttons made of plastic, glass, metal, leather, and even a few of wood. Old-fashioned buttons of bone for tiny doll clothes; shell buttons that had been punched out of Mississippi River clam-shells; sparkling round buttons jeweled with rhinestone; and large toggle buttons cut from a winter coat. Even a single gold-like metal button from a uniform.

Things got complicated when I realized most could be sorted into more than one category. Should a red four-holer go in the pile of red or four-holers? Did a black, fancy one go with fancy or black? Where should I put a blue, two-hole plastic button?

The afternoon wore on as I sat at the kitchen table, engrossed; I hadn't noticed rays of sun peeking through the west window. The rain had stopped. Muddy water flowing off the hillside onto the driveway had cleared. In Badger Creek, the water level was down and the creek had slowed. Dad suggested we have an early supper so he could finish milking and chores with time to spare before dark.

"Water 'n Badger's up high now, that'll bring the trout up-stream," he mused aloud. "Water's clearin' up, too, so they'll see my bait. Think I just might catch us a mess of trout this evenin'."

"And we'll have trout for breakfast," Mom added, brightly.

Just as well; I'd grown tired of sorting. When Mom asked me to clear and set the table, I happily jumbled the buttons all together and dumped them back into the button box, ready for another rainy day.

RURAL FREE DELIVERY

A Wisconsin farm never lacked chores: filling the wood box, pulling and hoeing weeds in the garden, feeding the calves, and the usual kitchen duties. Our parents expected us to do our part without "bellyaching." And we did. Mostly. Some jobs were better than others. Washing the mud-coated udders of a cow that had slogged through a swampy pasture was not as much fun as walking to the pasture to get the cows and drive them down the fragrant wooded lane toward the barn.

However, one chore was a form of escape: Sam and I loved getting the mail. Our mailbox stood at the side of Highway C, in the shelter of two towering white pines—trees remaining from the pine-era cutover at the turn of the twentieth century. Dad had anchored the oversized mailbox on a hand-cut post and cemented it into an old ten-gallon milk can. Mom had painted the mailbox pale green, artistically hand-lettering the block letters:

FREEMAN PRILAMAN
ROUTE 3
BOX 74

No one in our family ever considered driving the car the half-mile out and back to get the mail. Unless a trip to town was planned, we would either walk or bike to the mailbox.

Besides getting the mail, trips to the mailbox allowed us to keep an eye on the activities of our nearest neighbors. We usually brought home a daily report: Jerry's sheep got out of the pasture fence, or his bees swarmed again; we saw Albert repairing his fence line; Ole brought in a load of wood with his little Ford tractor. Sometimes we gathered the news with a sniff: fresh cut hay drying in the sun or being loaded on a wagon; newly plowed spring soil; the putrid odor of cow manure, ripe with age, after someone's spring barn cleaning.

On our return trips, we clutched spring's first pussy willows in one hand, the mail in the other. We knew exactly where the white trout lilies grew and when it was time to collect the green leeks of early spring. In May, we searched for yellow cowslips and snowy trilliums; in June, handfuls of tiny wild strawberries; July brought the first wild red raspberries that lined our road. In the fall, snakes of various kinds rested in rays of sunshine, collecting the radiating heat of earth from the warm gravel road: striped black and yellow garter snakes and tiny brown-backed, red-bellied snakes. In winter, we observed how snow driven into drifts became stunning works of wind-formed art. How could drifts extend so far out into the air, and yet, by some mysterious principle, become so firm and solid that Sam and I could walk on top of them? We observed the dynamics of weather, wind, and the powerful force of water. Each season presented a new, surprising lesson.

Shooing a green frog, a warty toad, or a covey of partridges from our path, all creatures we knew, was not uncommon. But once I hurried home and breathlessly summoned Mom to identify the mysterious bird on the road. She named it a nighthawk—a bird rarely seen in daylight, except this one had an injured wing. Another time, we raced home to beg Mom come see the white ratlike creature that lay so still beside the road. When we returned, the large animal had gone. We later realized it was a 'possum, uncommon in our area. It had played dead when we'd encountered it, then stealthily slipped away.

Most of our sightings were sanguine affairs of everyday events, though one day my return bike trip ended with quite a bang. I liked to zoom down the hill toward our house as fast as I could go while safely navigating the sharp turn over the bridge and into our yard. This day as others, I pedaled at top speed, until, just before my turn, a buzzing cloud of Jerry's bees swarmed in my path. I couldn't avoid them, so I went through them; bees hit my arms, face, hair . . . everywhere. Luckily, I sped so fast I escaped being stung.

Another mailbox moment topped the bee disturbance. On foot that day, Sam and I had reached the top of one of the rolling hills about halfway to the mailbox that offered a view of Highway C. "*Sam*! Look! There's a house movin' down the road!"

"Holy *cats*! Let's run 'n' see what's goin' on!"

Ed Halberg's "Cat," a yellow Caterpillar bulldozer, pulled a boxy, two-story house covered in gray bricklike siding. (People called Ed for numerous neighborhood jobs—whenever they needed to dig a pond, break a new road, clear some land, bulldoze a snowdrift. Or, apparently, move a house.) In this case, moving the house from its spot overlooking Deer Lake to Badger Creek, two bulldozers had been required to pull it up a hill. But once on Highway C, Ed moved it alone. Traveling up and over several miles of hills had challenged the dozer. When the house passed Sam and me, we saw massive timbers had been placed under it, instead of wheels. Folks said later that the rugged journey wore down those timbers to half their size.

Our hikes to the mailbox provided entertainment and feelings of accomplishment and, inadvertently, encouraged common sense and decision making. While cultivating the skills we needed for the future, the walks brought us a sense of adventure, encouraged our independence, and sharpened our powers of observation. Little did we realize we were doing much more than simply getting the mail.

PERFECT IN BLACK AND WHITE

The camera had been a gift from Mom's oldest brother, Vic, on the occasion of her high school graduation in 1930. The weighty and compact Kodak must have been an extravagant gift. She used it

sparingly during those difficult Depression days. Precious and few
black-and-white images recorded the trendy, '30s styles: Mom's
finger-waved hair under a close-fitting hat; heels, t-straps, and
pumps; dresses, slim and flared. Mom posing at the side of a heavy
black car of the era; friends playing ukuleles and singing on a
warm summer evening.

Eight years went by before Mom became the bride of her
country husband. Just one jaunty photo was taken on the Sunday
afternoon of the unannounced wedding. While Mom and Dad
built their modest home of logs, they were too busy and cash-
strapped to capture on film their exhausting work. Mom began
to take photos again when Sam appears outside our home as a
baby on a blanket. Then she took another baby photo—mine.

As we grew, Mom took photos of Sam's first toddling steps
with Smokey at his side; Sam beaming after catching his first big
fish; Sam's and my shivering and wading in Badger Creek; my
riding off for the first day of school on Sam's back bicycle fender;
my towing a sled in white winter wonder; my longsuffering cats.
And me and my favorite of all summer pastimes, mud pies.

Mom took family photos of both sets of my grandparents on
their fiftieth wedding anniversary celebrations and of Dad's parents holding bouquets of garden goods and flowers. On a summer's day, she snapped Dad in his sweaty work clothes, subtle
shades of black and white showing, bringing to mind his hard
work pitching hay on that steamy day. On another day, she took
one of Dad, his forest green canoe, and a prize muskie. Dad occasionally took photos: one of Mom with Sam and me—the three
of us showing off in our new Easter outfits.

I thought the camera was an amazing wonder housed in a
magical black box. I liked its bulk and how it folded neatly in and
out, opening and closing. I liked that it transformed moments
into memories. And I really liked that Mom let me use it. She
showed me how to unwrap the foil-covered film, carefully fit it

The Prilamans' fiftieth wedding anniversary, 1947

The Walhovds' fiftieth wedding anniversary, 1952

onto the sprockets, slowly wind until the first number appeared, and steady the camera while firmly clicking the shutter. She taught me to take care in selecting the right detail, the right shot, so as to not waste a single one. Lessons in patience and frugality. My first shaky attempts produced off-center, blurry images of my cats.

Being allowed to use the camera was a vote of confidence, for I knew that the cost of an orange-yellow box containing a roll of Kodak film represented another tiny cut from the monthly milk check. Developing the photos cost money, too. Mom and I removed film from the camera, sheltered it from the sunlight, carefully rolled it tightly, then placed it into the lumpy blue and white mailer and sent it to Ray's Photo Service in St. Paul, Minnesota. A week or so later, as we so often did, Sam and I pedaled our bicycles furiously to the mailbox hoping to find the fat envelope containing finished photos.

When I was nine, Mom sent away for a box camera for Christmas. For me! It was plain, square, and simple, only able to take black-and-white photos, yet it was mine. Not much later, we were amazed by our Indiana relatives' cameras with bright flashes— aluminum reflectors lighting up a dark room. Then instant color film and Polaroids appeared.

But for me, the old boxy black camera was good enough.

AT GRANDPA SAM AND GRANDMA BESSIE'S

Grandma Bessie and Grandpa Sam had neither car nor cow. We visited often to bring them groceries or milk from home. In winter, when they were unable to fetch water from Badger Creek, we brought them cream cans filled with water from our pump. We also served as chauffeur, picking them up for trips to the doctor and church in town and programs at Valley View schoolhouse. As they aged, we often stopped by just to check on their welfare. No matter the reason for our visit, Grandma Bessie always wanted us to stay for a meal. No matter how much company she'd had, she always said she was lonely. Thus, we rarely passed without stopping in as we went to and fro.

If we were on our way somewhere, Grandma Bessie looked us over carefully. She applauded Mom—at least in her diary—when she wrote, "Freeman's stopped in on the way to church dressed for Easter, looked so nice." ("Freeman's" described all four of us in our nuclear family.)

Sometimes Sam and I took refuge at our grandparents' during heavy rain on our walk home from school. At other times, Grandma Bessie took care of Shirley, Sam, me, or all of us. We three earned her approval by being able to entertain ourselves. Her diary notes with pride, "Kids played so nicely all afternoon." How could we not—with such a diversion as a packed attic?

We would climb the stairway to the yawning unfinished second floor. Already-read magazines and stacks of 78 rpm records sat on an ancient table beside a wind-up phonograph. We lifted the lid, placed the needle on a scratchy record, and cranked faster and faster to speed up the sound, laughing at the old-timey tunes. When we tired of going through the records and flipping worn pages of magazines, we took an awestruck look at Grandpa Sam's bicycle—the one he'd used to traverse the country for nearly two years, beginning in 1914. As far as we knew, no one had ever ridden it since it was placed in the attic four decades earlier. It would have been impossible to ride on the rough country roads, anyway, especially because no tires remained on the pieced-together brittle wooden rims. The wooden seat and strange-looking handle bars made it hard to picture Grandpa Sam on that bike.

Since Shirley lived just across the road, she often visited when her parents were doing demanding work or just to keep Grandma Bessie company. As the oldest grandchild, Sam sometimes walked down to carry in wood and water or help with household chores. Occasionally, he even delivered milk in a tin pail with a push-on lid and wire handle. Grandma Bessie thought it would be a good idea to give Sam piano lessons in return. Sam was more interested in coloring the pages of the Peter Rabbit piano book, and the idea was soon dismissed.

She never tried with me—even when she had me alone for dinner. But she did ask what I'd like to have. "Carrots with gravy," I'd reply. Grandma Bessie knew I meant her "white gravy" with sliced carrots. No matter what it was called—cream sauce, white sauce, white gravy—any veggie covered with the white substance made it more pleasant in my book: creamed peas, creamed cabbage, creamed onions, creamed green beans with bits of bacon (the best!). Along with meat gravy, what Mom mostly made, white sauce was a 1950s staple—practically one of the basic food groups. Grandma Bessie served it all in my favorite dishes: fluted, clear glass dime store dishes.

Grandma Bessie's small kitchen looked out on Badger Creek, with its back-porch entry facing toward the outhouse and where the old barn had been. She kept a water bucket and dipper near the door and a pantry just behind her kitchen. Like at our house, her cookstove kept the kitchen cozy. Her black cookie jar, decorated with a cluster of red cherries and usually filled with coconut bars, sat on top of the warming oven above the six-lidded stovetop, alongside the teapot that looked like an English cottage. Her tea kettle simmered lazily at the back of the stove, always, unless it was summer.

Most of the activity in the home centered in the dining/living room, around a long dining room table surrounded by six chairs. A large framed photo hung on the wall—circa 1913—taken in Canada, just before the family's return to Indiana. In it, four-year-old Dad wears all white, except for classic Buster Brown shoes and a black belt dipping low in front, which appears to sport a metallic crest of arms. A ring on Dad's left middle finger seems to gleam as much as the crest and the belt buckle on the side. The look on

Dad's face possibly registers resignation regarding his all-white ensemble—a dress, with starched broad collars trimmed with lace and long puffy sleeves cuffed with lace, knee-length pantaloons peeking out from below, the bottom portion of his legs covered with white stockings. Dad stands on a chair and towers over his ten-year-old brother. In contrast, Jerry wears all black, except for a jaunty white hankie poking out from his double-breasted suit pocket and his crisp white shirt, shiny necktie at his throat. He, too, wears short pants and a ring on his middle finger, albeit on the right, and the look on his face seems pleading: Are we done yet? I spent what seemed like hours contemplating that photo in puzzlement—neither boy seemed anything like the lumberjacks I'd come to know.

Grandpa Sam's Boston rocker and his cot, where he took a nap each afternoon, sat near the photo. He slept through the chiming

Grandpa Sam in his favorite rocker

clock standing on a knit doily on its own little shelf in the corner
of the room, as the pendulum swung and struck the hours. Other
armchairs circled the dining/living room and sometimes spilled
into the adjacent parlor.

My grandparents kept the wood furnace in the basement
stuffed with fuel from their novel woodshed. When they'd built
the home, they left the last large room off the hallway unheated
and unfinished, down to its dirt floor. A hallway door opened to
their convenient woodshed, as if it were just another bedroom,
right next to the stairs heading to the basement. Despite that
wood furnace, the far corners of the rambling house were drafty
and cold.

In winter, my grandparents blocked off the parlor and its
connected extra bedroom by hanging a Canadian wool blanket
over the doorway. Mom and Dad had stayed in that chilly extra
bedroom for the two years after their marriage while they built
our log home. No doubt Grandma Bessie had made them com-
fortable then as she did with guests now: placing a freshly made
straw-tick "mattress" on the bedsprings and covering it with fluffy
and soft feather-tick—a sort of comforter stuffed full of feathers
from her own chickens.

But in summer, well, that was different altogether. In summertime, on the long porch preceding the entryway overlooking
peaceful Badger Creek tumbling below, relatives, friends, and
neighbors would sit outside on benches to visit.

And inside, the house took on a breezy air, especially in the
parlor. A green leafy canopy first shaded and cooled the air, which
wafted through open windows. Dappled afternoon sun flitted into
the rambling white frame house. The gauzy curtains on Grandma
Bessie's parlor window would lift and I would smell summer on
the wind as I surveyed my surroundings. The proper parlor contained Grandma Bessie's red velvet settee—its arms and back
covered with starched, crocheted white doilies. Her parlor tables

were covered with embroidered white dresser scarves (table runners), homemade lace around their edges. But the pièce de résis-
tance was the black upright piano.

When I was very small, Grandma Bessie took her turn hosting worship services—home church—pounding out hymn after
hymn for the ladies of the neighborhood. Many women walked
across field and forest to share Sunday church in her parlor. A few
came from Exeland, driven by their husbands. Men in dark suits
and vests, white shirts and ties, arrived with their wives in black
cars. The men sauntered off with Grandpa Sam to explore the
garden, while the women surrounded the upright piano, donned

Grandma Bessie's piano and "home church"

in their Sunday-best dresses, heavy and draping, in muted purple tones and paisley design. While the women raised their once-clear voices in song, an occasional voice straying with age yet its message no less sincere, I would stare at their blocky, high-laced leather shoes—black and ancient. To me, as ancient as the women themselves. These women, mostly pioneer women, would tie a knot tightly around small coins in the corner of a hankie for me to deposit in the wooden collection plate when it came around. And when my birthday came around, they would tuck a colorful handkerchief folded into a birthday card for me.

Beside the piano and bookcase, Grandma Bessie kept a T. Eaton Tea tin from Winnipeg containing dozens of one-cent photo and picture postcards with cheery holiday messages beautifully embossed in color with detailed designs. I would come to know later that each backside contained handwritten messages of longing, loneliness, celebration, or encouragement. They'd been sent among family, Canada to Indiana and back, during Grandma Bessie's absent years—1905 to 1913. Much of the penmanship was perfect, but some notes were hurriedly scrawled, back in the day when people knew folks by their handwriting. Christmas cards, anniversary cards, and birthday cards, some addressed to "Master Freeman Prilaman."

Finally, when Mom gave the all-clear, I could play with the noisier diversions: the stereoscope viewer cards in 3-D, which were relatively silent except for the removal, placement, and sliding of cards into focus in the wire holder; a wooden Carrom board, with sticks and shooters; Dad's childhood marble collection—some made of clay-like crockery, as well as the glass-ies, cat eyes, solids, swirlies, and gleaming metal steelies. And then, home church ended for the day—and shortly thereafter, for all time, replaced by church in town.

It remains just as precious a memory as that of the hours and hours I spent at Grandpa Sam and Grandma Bessie's—in the

kitchen, in the attic, and in the parlor, with the strains of music
and the melody of pioneering women's voices singing together
in a hymn of harmony.

DEER LAKE WITH DAD

The commandment on the plaster plaque hanging on our kitchen
wall, the plaque I'd painted in vacation bible school, said: "Remember the Sabbath and keep it Holy." It said nothing about
going fishing.

Neighbors claimed they could set their clocks by Dad. When
he finished his chores, he gathered the thermos, the lunch Mom
had packed, and Mom to help him lift up the forest green canoe
from its perch behind the house onto an ugly wooden rack permanently clamped with suction cups to the top of our '49 Ford.
The rack, the canoe, the car, or all three, identified our family so
everyone knew everywhere we went. None of this mattered to
Dad. Placing canoe paddles in the trunk, he happily took off on
his fishing quest, bumping down roads to lakes unknown. Often,
Rob, Ole, or Jerry joined him. Dad valued his privacy and places
known only to himself and his three trusted partners. Once, he
had invited a new neighbor who claimed to be an avid fisher-
man and took him to a favorite haunt. Later, word of Dad's secret
fishing hole had spread all over town, violating the fishermen's
unspoken rule.

Haymaking was the order of one summer's day in July, but
after the previous night's rain, it was too wet to mow. When
the gray and overcast sky had not cleared by midmorning, Dad
earned a day of fishing. He walked with his shovel to the site of
our old chicken coop, while I got an empty coffee can from the
junk heap. Dad turned over brown clumps of earth, and I picked
up the angleworms and a few big night crawlers, plunking them
into the can.

Mom helped Dad get the canoe on the car, toss tie-down ropes over the top, and fasten them. With paddles, float cushions, and a life jacket for me, Dad and I were off!

The canoe was one of Dad's valued possessions, mail-ordered from Sears—some assembly required. Dad had carefully pieced together and finished the items included in the kit: wooden gunwales, curved white cedar ribs, strips and rails of spruce, and caned seats; canvas, seam filler, varnish, brass fasteners, and the trademark marine finish in dark green. Carved wooden paddles completed the kit. Dad's sleek, durable, and transportable canoe provided not only hundreds of pounds of fresh fish for the family table, but hours of peaceful diversion and a way to answer the invitation of thousands of miles of serene watery wilderness. The well-traveled canoe bore more history than words could record; it had logged hundreds of water miles. It had jolted over nearly impassable roads to lakes unknown to most fishermen.

But on this day, Dad and I traveled to Deer Lake, nearest our home, just a couple of miles away. As our car left Highway C and bounced down the gravel Deer Lake Road, the canoe shifted in its straps. Swinging around mud holes filled with brown water, we passed a long-time hunting camp in the woods and the little clearing, where we eased past a vacant tiny house, previously inhabited by a now-deceased bachelor. Few residents had ever lived in this area. The ornery hermit of Dad's lumber camp lore had once lived deep in the woods a few miles south of the lake, but he was long gone; and the area's remote Lutheran retreat for wayward boys was set so far back we wouldn't be able to tell if it was full or empty. But we knew it was usually empty.

Shrouded in the late morning fog of the vast Rusk County Forest, relative wilderness surrounded the lake. The uninterrupted woods never failed to spawn an atmosphere of mystery. There were no beaches; no one ever swam at strange Deer Lake. They weren't brave enough. For Deer Lake offered the intrigue of a quaking bog—floating islands of peat and sphagnum moss, called muskeg. Pieces of the bog broke away and moved with the wind. Sometimes they lodged against the landing, making canoe entry impossible. In strong winds they sailed the lake from end to end. The previous night's winds had been gusty, shifting the bog to the far shore; the landing at the base of the hill on Deer Lake was open. While Dad prepared his gear for launch, I scanned the area for snakes, knowing they were routinely present. Lots of them. Too darn many.

We considered Deer Lake "our" lake. Although puzzling, it held no terror—at least for Dad, who knew its moods. The lake itself was dark with tannins and decayed vegetation and covered by the boggy layer of muskeg, which supported sundew and pitcher plants. These carnivorous plants further enhanced the lake's mysterious image. Sundews snared insects in a sticky substance after insects crawled in; pitchers, gape-mouthed and blood red, lured insects in and clamped their mouths shut. People courageous enough to set foot on the bog found a robust and hungry crop of black flies, mosquitoes, and blood-sucking wood ticks. Those few who explored the spongy bog could find themselves sinking through up to their armpits in the watery unknown or face-to-face with a curious black snake. Long water snakes were not for the faint of heart. Deer Lake had a large and healthy population of harmless reptiles that created a scare when sliding in search of a frog on a lily pad or moving alongside our canoe for a better look. Some could get mighty big.

We slipped away from the landing, our canoe gliding through pitch-black water, our eyes scanning for signs of movement. Blue herons stood still and silent on overhanging snags waiting to spear a fish for dinner. I peered into the lakeshore looking for a break in the thick vegetation where I might catch a glimpse of a doe leading her young red fawns for a drink. Dad paddled to the north end, near the base of the huge Meteor Hill.

I think Dad was happy for my company, but mostly he just wanted to fish. Real fishermen don't talk much. We chatted now and then. Otherwise, we remained silent. I watched him tie lures, spoons, rubber worms, and wiggling earthworms to his hook. Dad targeted his casts near a downed log, an overhanging branch, or a shore overhang, all likely fish haunts. His line whirred as it sailed by, landing with a *splunk* in the gloomy water. Sometimes he changed his rod, reel, or bait, and occasionally he landed a fish. Dad rustled noisily through his metal tackle box, then reached for his pocketknife to cut the line and used pliers to remove the hook. He threaded a white rope through the gills and dropped the fish into the bucket if it was obviously legal size. If questionable, he measured it by the yardstick fastened on the bottom of the canoe and weighed it on a scale. Dad released small, squirming fish with a *plunk* into the water. As a fish realized its freedom, we watched it swim hurriedly away. Sometimes the lure or line snagged on a log or tangled in brush. Then Dad put his rod aside and with a single paddle easily maneuvered the canoe to the site of the problem.

I listened for the sounds of disturbed herons, distant ravens, and the caws of raucous crows. No other people were anywhere in sight. The faraway bark of a dog, unidentified bird noises, and water droplets off Dad's canoe paddle were, for a long time, the only sounds besides Dad's whirring lines and water lazily lapping against the canoe. I dipped my hands into the inky water, wondering aloud, "Dad, could a muskie grab my fingers?"

"Sure could." My fingers jerked up quickly.

The day soon became long and tiresome; I wished we could just go home and wiggled like a fish on the line. Dad tried to entertain me, paddling closer to the bog to take a better look and explaining about the peculiar plants as we drifted. We paddled up the watery channel that flowed into the lake, where we watched a buck drink water beside a tipping willow. Herons lifted from their fishing perches. An osprey circled in search of his next meal. A partridge drummed somewhere on shore. Water droplets skittered across a waxy lily pad in the breeze. I picked white and yellow water lilies, tucking them into the gunnels of our cedar canoe.

Still, I had grown tired of sitting.

My spirits soared when Dad's eyes raised to the sun in the sky, gauging the time of day just as he did in the hayfield. The sun had broken through the clouds, beating hot now; the sky had completely cleared. He pulled from his overalls the leather strap that attached to his pocket watch.

"Yep, the afternoon's gettin' on," he murmured to himself. (Great!) Then, "Just one more trip around the north end 'n' we'll head for home." It was not the news I wanted to hear.

Finally, our canoe glided to the landing, we lifted it to the roof of the car, snapped the ropes in place, and drove on home, with a few bass on the string. When he handed them to Mom, she said, as she so often did, "Sure wish you'd learn how to clean fish."

(Dad had known how to clean fish for nearly forty years.)

"Nope," Dad teased. "Not unless you get the cows in 'n' get ready for the night's milkin.'"

SUNDAY SOJOURNS WITH MOM

Twice a month on Sundays, we saw Grandpa and Grandma Walhovd in Birchwood; after visiting with his in-laws for a bit, Dad would slip away to fish. But on the other Sundays, Dad left from our house and Mom, Sam, and I were free for a day of our own.

Mom enjoyed the pace of a day without her usual cooking duties and extra chores. Even before we were born, Mom had walked over hill and dale to visit with neighbors whenever she had free time. She had been raised a town girl, with seven siblings, who then had to seek out social opportunities as she adjusted to

Mom on a Sunday sojourn

her new farm neighborhood. Now we looked forward to those ranging visits, too. We roved about as the crow flies, taking the shortest distance, as we explored nearly lost paths and often followed Badger Creek.

Sometimes we walked to the west, down the cow lane, across the pasture, and through the woods to visit the Clarks. The Clark kids attended Meteor School, not Valley View as we did, though they lived just a couple forty-acre plots away. But we knew them well as neighbors; Dad happily hired one of the boys when he needed an extra hand. He welcomed their help and expected they welcomed the extra income. Some of the ten Clark kids had grown up and moved away, but some of the younger kids, two close to my age, were sure to be at home on Sundays.

Mom admired her neighbor, declaring, "I think the world of Mrs. Clark." She and Mom were of similar age, but Mom never called her by her first name, Retta. Mrs. Clark was a tiny, frail-looking woman; quiet and soft-spoken, with a kind and humble heart. She cared for the family by herself, since her husband drove to work each week in Illinois, over 350 miles away. Mom thought Mrs. Clark must feel isolated in the sole care of the big family, especially since she did not drive.

Mom seemed to return home renewed after a chat with her farm neighbor, hoping Mrs. Clark felt the same. One Sunday, after Mom had carried a loaf of bread over in her apron, Mrs. Clark found a way to return the favor. When we went to bring in the cows from the pasture on the next afternoon, we found the bucket of freshly picked wild raspberries she'd left at our pasture gate; a token of thanks for the neighborly exchange.

On the walk home from the Clarks', we always located our cows; they sometimes gathered in the far corner of the pasture. Sam guided them down the winding lane toward the barnyard for evening milking. If it were June, Mom and I scanned the cattail marsh where I stopped to search for my favorite June wildflower growing near the pasture gate: blue flag iris. They seemed to wave cheerfully, displaying a jaunty welcome to early summer while flourishing in swampy muck. Our cows tromped through the same mud hole, even though they could have gone around—but cows don't do that. Their hooves made sucking sounds when they yanked their coated ankles out onto firm ground. Somehow, they avoided the patch of blooming iris.

My favorite destination every other Sunday was the old Mulkey place. A fence line still separated what had been the Mulkey land from Dad's first forty, but another tumbling tributary of Badger Creek drew the forties together. The stream meandered through an enchanting ravine, winding and flowing on past our house, the same ravine where I saw a fantasyland of horse thieves and Apaches peeking around trees and rocks.

If we went to the Mulkey place in late May, our mission would be to explore the remaining portion of the log house. It sat beside a water source, the spring that trickled into the Badger Creek tributary. We peered through a sagging windowsill into the ruins, just enough for me to picture the folks who once lived there. A broken glass jar and an old leather shoe still lay on the floor. We tramped curiously about, always on the lookout for the long, spotted pine snakes that so often frequented old log dwellings. Then we went to retrieve the real prize of our visit: A wondrous patch of red rhubarb—what Dad and old-timers called the "pie plant"—produced the thickest stalks I'd ever seen. Mom yanked the stems from the ground, removed the gigantic, poisonous leaves, and dropped the stalks into her packsack. Sam wandered a short distance away, following his nose for the sweet scent of lilacs. With his jackknife, he sliced off full branches for a bouquet to brighten our table.

"It's a double bonus day!" Mom exclaimed. We agreed. And I knew we would have another when we returned in a few days with blue two-quart canning jars, gathering more lilacs to place on family gravesites for Decoration Day (Memorial Day), just around the corner.

On one June Sunday, our walk took us across the hayfield behind our barn and through the woods to the farm of Lena and August Erickson, neighbors bordering the north side of our property. On the way, we passed more than one giant rock pile, stopping to look for the pink wild roses that grew only there. "Those wild roses are getting to be scarce as hen's teeth these days," Mom commented.

As we meandered on, we kept our eyes peeled for other plants, too. Mom had a special knack for spotting four-leaf clovers as we walked through the fields. We took them home and pressed them, for good luck. Try as we might, Sam and I could never come close to matching her skill.

Then we combed one tiny corner of the woods, the only place we knew where we might find a rare yellow moccasin flower, before moving cautiously through the barbed-wire fence that divided the copse from Erickson's open pasture. At the edge of the woods, we took note of the pasture's occupants. Most often, calves, heifers, and cows lifted their heads curiously in our direction and then went back to eating. But we applied country wisdom, noting whether or not August Erickson had left the bull in the pasture to woo his lady companions. Hearing no bellowing calls summoning them to his side, we proceeded safely toward the farmyard, then paused again as we neared Lena's yard. A ruckus of cackling geese, gobbling turkeys, quacking ducks, and a handful of noisy chickens announced our approach. Most were simply curious, but the long-necked, orange-beaked geese headed straight toward us, necks outstretched, beaks eager to clamp, pinch, and twist a bit of tender skin. The commotion in the farmyard alerted Lena; she emerged onto her porch waving a dishtowel in greeting. On previous visits, we'd learned to await safe escort. She corralled the hissing geese, calmed the turkeys, and sent the dog to herd the ducks and chickens so we could enter.

Once in the safety of her kitchen, Lena dried her hands on her apron, put aside her potato peeling, and directed us toward the darkened parlor. Clearly happy to see us, she dropped heavily into a chair for a visit. In her rich, full voice, Lena chattered on and made inquiries; Mom answered. When I decided the conversation was of little interest, I gazed at the ivy-leaf wallpaper in the dining room and deep maroon drapes keeping the room dark, until Lena offered tea and cookies.

When it was time to go, Lena requested a moment of prayer, making me feel slightly uncomfortable. Why did we need to be prayed over in the middle of an ordinary Sunday afternoon? Mom's polite compliance signaled we would do the same. It wasn't clear to me if the prayer was on our behalf or Lena's. I decided she did it for all of her guests. Mom thanked her for the hospitality, visit, and prayer. Lena seemed pleased to send us on our blessed way. She guided the safe passage of our departure though the menagerie of fowl, then waved a dishtowel goodbye across the field as we walked toward home. Nearing the woods, I demanded we stop to survey the site where Dad had shown me a bachelor's long-vanished cabin. Only the slight depression where the cabin had once stood and the smooth, flat stepping stone marked the imagined entryway. But we couldn't linger. Mom had brought Lena a jar of jam, and now Mom carried Lena's slip of a plant rolled in damp newspaper, tucked into her apron; the cutting would root in water and become a new African violet.

Our Sunday escapes took us on visits in nearly every direction: west across our pasture and through the woods, north through the hayfield and other woods, and east, down Badger Creek, along the tote road, to Grandpa Sam and Grandma Bessie's garden, aged apple orchard, and house. Or to Jerry, Millie, and Shirley's, across the road from my grandparents. Sometimes we ranged as far as Deer Lake or to Rob and Inee's near Valley View School. Having no telephone to warn, possibly we'd arrive to find our intended target absent, but there was likely someone home just a bit farther on. Either that, or we'd turn around and take a leisurely, explorative walk back home. Sunday sojourns with Mom seemed a fine way to end one week and be ready to launch refreshed into another.

But sometimes the weather didn't cooperate. On drizzly Sundays, we stayed at home. Mom took a rare afternoon nap or sewed and sometimes cooked up a special recipe, something she'd learned to make while working at Garbutt's Island Resort before she married Dad—baked Alaska or floating island. Then she read to us, told us her childhood stories, and taught us rhymes and songs: "Two Little Blackbirds," "Old Dan Tucker," and one of my favorites, especially on a rainy day, "It Ain't Gonna Rain No More." I paraded around the kitchen belting out the chorus and giggling at the query:

It ain't gonna rain no more, no more, it ain't gonna rain
  no more
How in the heck kin I wash my neck if it ain't gonna rain
  no more?

When Mom taught us the 1894 song "Oh Playmate, Come Out and Play with Me" by Philip Wingate and H. W. Petrie, the refrain reminded me of visits to Grandma and Grandpa Walhovd, their property boasting, as it did, rain barrels, apple trees, and a cellar door:

I don't want to play in your yard,
I don't like you anymore,
You'll be sorry when you see me,
Sliding down our cellar door,
You can't holler down our rain barrel,
You can't climb our apple tree,
I don't want to play in your yard,
If you won't be good to me.

Rain or shine, as evening neared, we looked forward to Dad's return to see what fish he'd caught or bird he'd shot, during fall hunting season. We looked on while Dad pulled colorful duck or partridge fantail feathers and then examined the contents of a bird's gizzard, to see what it had been eating. Often, it was red partridgeberries.

Dad returned with morel mushrooms in early spring, yellow or white waterlilies in summer, and wild cranberries in fall. Once he brought an unusual fringed orchid he'd found. Another time, he gave me perfect gray pottery clay he'd dug from a marsh; I'd shape and bake it in the sun. Once he brought a pine knot. He told Sam and me to wait until dark when it would glow green with natural phosphorescence. It did!

Over some Sunday evening suppers in May, Dad told of the
big bass that got away when the line tangled on a downed log,
and we told Dad of our visits to the Clarks and the old Mulkey
log house. Luckily, there had been enough time for Mom: in the
fragrance of fresh lilacs, we all enjoyed fresh rhubarb pie.

DRIVING LESSONS

Mom didn't drive. Most women in the neighborhood, including Ruth, Daisy, Inee, Lois, and Della, didn't, either. Husbands
drove wives where they thought women needed to go, such as
to the Meteor schoolhouse. Farmwives often gathered there for
monthly meetings of the Meteor Homemakers Club, which occasionally featured a special event conducted by a county home
extension agent.

Though it was likely that few in our community were familiar
with the term "The Wisconsin Idea," referring to the University
of Wisconsin in Madison reaching out to help in all corners of
the state, it was alive and well and operated in the form of county
agricultural extension agents stationed in each county seat. Our
nearest agents were located in the Sawyer County Courthouse
in Hayward, but they also came visiting to assist farmers and
homemakers. The former benefited from soil testing and the latest information on fertilization, crop rotation, and animal care.
For the latter, agents gave advice and provided how-to lessons in
economical living: canning, gardening, cooking, cleaning, and
sewing. Agents presented new ideas as they came on the scene.
Many women, like Mom, engaged in a sudden surge of 1950s
creativity—making hammered-tin serving bowls and trays and
aluminum drinking glasses, along with the equally popular woven
reed baskets.

One snowy winter's day, the Sawyer County extension agent,
who was running a tad late, hurriedly drove down the steep

Meteor Hill en route to the schoolhouse. She slowed for the turn at the bottom, but not before hitting an icy patch, causing her car to suddenly lurch sideways and spill the objects of her presentation to the floor. Fortunately, her quick action kept her from sliding all the way into the deep ditch, and she arrived minutes later, safe but a bit shaken. Unfortunately, she discovered the destruction of most of what she'd brought for a demonstration on the ease of preparation, the nutritional value, and the many uses of eggs. Most of her three dozen had cracked and turned into a yolk-eggshell mix. Instead of culminating her lesson with Golden Yellow Sunshine Cake, she abruptly changed to the nutritious benefits of scrambled egg bakes, omelets, and soufflés. No one knew the difference until she began the car cleanup job before her drive back to Hayward. The ladies of Meteor provided help—and loved to recount the story of the disastrous drive of the woman from the Sawyer County seat.

Of course, there were a few local female-driver exceptions. Bernice Fairman drove; she and Grandma Fairman prepared and delivered the hot lunch to our country school each day. Later, Bernice drove to her job at the bank in Bruce. She'd learned to drive in Chicago before moving to the country. Now that Flossie (Verna's mom) solely supported her three girls, she, too, drove to her work at the dentist's office. Soon Nola also had a job in town. It was, after all, the 1950s.

Change wafted through the air. The days of getting a driver's license by simply going to the post office or the local constable were over. Now, state certification required a driver's test with an examiner. Roads and cars were rapidly improving. America was on the move. Highway 48 had until recently been spewing clouds of dust on the unpaved road to Birchwood. Blacktop now replaced the gravel that had been as rough and bumpy as Mom's washboard. On the neighbors' television—we didn't own one— we watched President Eisenhower calling for his "Grand Plan" for speedy, safe travel on the vast interstate system of roads he hoped would soon crisscross the nation.

Mom took a notion that she would learn to drive. Dad stayed silent on the issue—whether with resignation or tacit approval, we didn't know. He did have to agree that it would be handy for Mom to be able to run to town to get parts for the tractor or feed from the mill. On occasion.

But I wondered. With a driver's license, would Mom be driving away on shopping trips as some neighbor women did? My mom wasn't the type to be going on afternoon visits any time she pleased—was she?

Soon we found Mom studying a driver's education booklet with its rules of the road—poring over signs and illustrations, the meaning of solid and broken white lines and solid yellow lines. When company came, she even asked questions about passing, parking, and turning around. One morning over cereal, Mom entertained us with the dream she'd had—proudly driving her Dodge Royal, a shiny pink convertible brandished with exaggerated high-swept wings, just as the radio ads described.

Ads also boasted of "triple turbine takeoffs," "eager acceleration," and "gliding rides." Just imagine blissfully rolling away, swiftly driving without a care on the open road as you tour scenic America in Dinah Shore's motorcar while her charming voice croons, encouraging folks to see the "greatest land of all"—"See the USA in your Chevrolet, America is asking you to call!"

Who could resist? Yes—Mom was going to drive!

Sam and I weren't invited to ride along on Mom's driving lessons. Our dead-end road was perfect for practice. We watched Mom's car bounce away as the '49 Ford jerked over the bridge, strangely with Dad sitting in the passenger seat. Then we collapsed in laughter as it lurched up the hill and out of sight, headed to the mailbox. But our jaws dropped when the car returned and Mom rolled smoothly to a stop.

Sam and I dreamed of leisurely Sunday afternoon drives, trips to Windfall Lake to swim, shopping trips to Rice Lake, and going to free outdoor movie nights in Exeland every week. But that summer eased along pretty much as every summer before, except for Mom's two "getaway" moments.

First, Mom announced she wanted to attend the Homemakers Achievement Day. It would be held twenty miles away at the hall in Stone Lake, a gathering of Sawyer County homemakers clubs to show off their current projects or crafts—like basket weaving and metal pounding. Mom planned to pick up Daisy, Ruth, Inee, and Millie—the non-drivers. One morning after finishing the breakfast dishes, she drove away and left us alone with Dad—and instructions for lunch. Dad worked on the farm as usual; without Mom there, Sam and I weren't sure what to do. Everything seemed strange, and the day dragged on forever while we waited and listened for the sound of our approaching car. Finally, we heard the crunch of gravel. Mom cruised to a stop as we rushed to hear the events of her day.

The next odd event came when Mom leaped at the opportunity to do the yearly cleaning of both Valley View and Meteor schoolhouses. The school board paid a local housewife fifty dollars for each, and this year Mom was thrilled to be awarded the job.

One summer morning, Sam and I drove off with Mom, wielding buckets, mops, scrub rags, and bright orange boxes of Spic and Span. Like at Valley View, I entered Meteor schoolhouse joyously, drinking in the smell of chalk and blackboard dust, red oily sweeping compound, lead pencil shavings, and books—pages *and* ink. The scent of years of accumulated schoolroom odors and the many children who had romped across those weathered wooden floors smelled just like Valley View. We sprang into action, slinging soapy water and sponges, helping Mom wash desks, scrub walls, and dust corners.

Over the course of several days, I was in heaven exploring every cranny of both schoolhouses in the summertime. Mom's newfound freedom had enabled me to snuggle into a corner be-hind the teacher's desk for the best treasure of all: reading every book unavailable to me on Valley View's three-shelf library.

My dad fishing

School Days at Valley View

VALLEY VIEW SCHOOLHOUSE:
ONE ROOM FOR ALL

Valley View School opened about the time my father, then seven years old, arrived in the Northwoods. Over three decades later, my school days would begin there.

Peggy's first grade class at Valley View, 1953–54. Peggy is in the second row, second from left.

One bright August morning, I climbed aboard the battered back fender of Sam's red bicycle, excited to be on my way to first grade as a five-year-old (country schools had no kindergarten). I clutched his waist tightly as he peddled down the dusty gravel road, too excited to notice Mom's worried look as she snapped a last-minute photo. We were off on the three-mile ride to Valley View School!

Ten-year-old Sam introduced me to our new teacher with an inauspicious, "This's my little sister, Peggy. She don't talk much."

Desks of wood and wrought iron, arranged in straight rows, lined my country classroom: seating for nearly thirty students, ages five to thirteen. First graders sat at small desks next to the long blackboard on the west wall, while the eighth graders occupied larger desks, looking out the east-facing windows toward Valley View Road.

Although eager to attend first grade, I soon became disenchanted. It was overwhelming to be away from home all day. Filled with needless worry and homesickness, I cried. My teacher, Miss Olson, allowed Sam to take me home at noon, which meant he'd miss afternoon schooling, as well. However, my adjustment to a full day of school improved by the time I turned six in October. Still, I struggled to write between the lines; I struggled to control the fat, unruly pencil that consistently smudged my crookedly formed block letters.

After my birthday, I happily wore my new favorite outfit: denim bib overalls. Mom had ordered them in pale red for me, from Wards. Either I was too shy to ask or I waited too long, but to my dismay, a wide puddle formed under my desk, growing larger by the second. Worse yet, the lower portion of my bibs rapidly turned bright crimson, matching my face and announcing to all that I was too late to go to the outdoor toilet. Miss Olson, again, allowed Sam to take me home. I hoped everyone would soon forget.

Finally adjusted to a full day, the next torture came in the form of a dreadful green phonics book. Putting parts of words together by sound mystified me. I struggled to hear the sounds. Why did I have to learn *parts* of words? When Mom read books and stories to me, the words told a story. Decoding words did not work for me. Fortunately, Miss Olson's up-to-date education launched me into reading when she introduced *Fun with Dick and Jane*.

Learning whole words and reading the simple stories in the first-grade reading circle motivated me. In those primers I met Dick, Jane, Sally, Father, Mother, Spot the dog, and Puff the cat.

Oh look, look, see Puff.
Oh look, look, see Spot.
Oh look, see Puff, see Spot.
Oh look, see Peggy read.

Off I soared with my new friends, down the road to reading in Father and Mother's green coupe, leaving the horrible green phonics book in the dusty distance.

COUNTRY SCHOOL DAYS
THROUGH THE SEASONS

We lived nearly as close to neighboring Meteor School as we did to Valley View—smack dab between the two country schools in the Town of Meteor; each rural Wisconsin school had been plotted so that no one lived more than three miles away. Getting to school was a matter of biking, sledding, walking, or running as we picked up the neighbor kids along the way, every day a new adventure.

After we'd traveled the distance of two forty-acre plots to the end of our dead-end road, Sam and I greeted Eric, Sonja, and, when I was in second grade, Johnny Halberg. Turning onto

Highway C and heading up over Carlson Hill a mile beyond, we gathered their cousins, Richard and Neal (Carol was still too young for school). When we reached the home on the corner belonging to Harry and Mina Whyte, we turned onto Valley View Road. There we met sisters Jane and Verna—or "Verna *Mae!*" if her mother was upset with her—then Bobby and Sandy. At the bottom of the steep Prilaman hill, we crossed over one of the branches of Badger Creek, next to Grandpa Sam and Grandma Bessie's house. Just across the road stood Jerry and Millie's log house, where we met Shirley and, in fall, picked the dry, pebbly-rough pods of milkweed on windy days and watched the silky seeds disperse like parachutes. Roger—Jane and Verna's cousin—joined us next, near the bridge over the main branch of Badger Creek; then came the farm only a mile away from school, where we picked up Nancy Birdsill.

We never crossed that beckoning bridge without stopping to look for trout dashing for cover or going under the bridge to catch skittering water striders, slimy frogs, and warty toads. It was easy to lose track of time. The resounding clangs of the early warning school bell jolted us into action. No time to waste! We panicked, racing breathlessly, our hearts pumping hard over the last mile to school. Gasping, sides aching, the Valley View Road kids—at least a quarter of the entire school—finally joined our friends and slipped wearily into our seats just as the bell rang its final bong.

Winter travel to school presented a challenge no less alluring, but of a different nature. When the gravel roads became snow covered and icy, it was time to determine who owned the finest sled. We learned sled science by close observation. Riding face down on our bellies, and heaped three or four high, we quickly learned that more weight meant faster travel. Small sleds seemed to have the speed advantage. But older, much-used sleds had the advantage of quicker and looser steering. Whether it was a Silver Streak or a Flexible Flyer didn't matter, as long as it was the fastest in our fleet and able to zoom down country roads.

Competitors examined their metal runners to make sure no rust had formed on them after a few days of disuse, and waxed them with bits of paraffin from their mother's jelly jar for increased glide and speed, before converging at the scene of the challenge, Prilaman hill. The sled that carried its riders the greatest distance would be proclaimed the winner. Clutching the handles on the front, the driver did his or her best to avoid hitting patches of gravel on the road, which slowed the descent; inevitably, we would watch sparks fly from the surface of the metal runners when the driver failed. Kids in our neighborhood deemed Flexible Flyers the Cadillac of sleds, as most years they were the champions. Though short in length, they could be piled four kids high on a belly run and they were lightning fast.

Once we chose Top Sled, it had the honor of being the *only* sled we used to get to school. But before launching, we reviewed the safety rules. Most important: steering. Learning to shift the sled's position quickly, so as to avoid an approaching car or truck, was an important safety skill. Next: *emergency* steering. The toes of our overshoes, usually the black-rubber, five-buckle kind, controlled emergency steering. Most of our overshoes had numerous rips and tears, which had been mended at home using a bicycle tire patch kit. After roughing up the boot with a metal scratcher, we'd cut a patch from an old inner tube, affix it with smelly glue, and hold tight until it dried. The patches soon tore off—no wonder. Dragging a boot for emergencies was not popular with parents, but it worked amazingly well. Last, and closely related to emergency steering: emergency stopping. A loud voice was also important to safety. When the driver shouted, "Bail off!" the pile of kids knew to roll off immediately—danger was imminent! Hazards included hitting an icy snowbank head first or careening off the bridge.

The smallest kid had the biggest challenge, hanging on for dear life while teetering at the top of the stack. He or she also had to push the sled to get it started with kids aboard, trying to make it move as fast as possible before leaping on. Miscalculation resulted in missing the ride entirely and watching everyone else hoot 'n holler to the bottom of the hill. For two years, I had that duty and grew tired of clinging to the top of the heap. What a relief when Johnny replaced me as the smallest kid.

When the roads turned especially icy, we set new sled distance records, measured by the number of electric poles passed. Then we would watch anxiously out the east bank of school windows, listening for the unwelcome sound of the Town of Meteor truck spreading sand to combat deepening ice. We knew then that our speed-sledding fun would be over and for a while we'd even have to walk to school.

I suppose snow made our three-mile trek more difficult, though we saw it only as more interesting. We tested walking on the crusts that formed on top of drifts, falling through to our waists and laughing while our friends pulled us out. Sometimes snow drifted so hard and deep that only Ed Halberg's bulldozer was powerful enough to open it up. We teetered atop the peaks of tall banks after climbing the increasing heights lining the road-side. Balancing on top while walking, it seemed as if we could nearly touch the electric wires hanging from the poles above. The shrill whine of the wires grew louder when temperatures dropped. The bitter morning cold caused trees to crack with ear-splitting pops, reports loud as gunshots, as subzero temps were falling even lower in the predawn darkness.

I don't know what criteria Mom used to decide when it was just too cold to make the trip to school, but on dangerously cold days, she kept us at home. Early each morning, she checked our thermometer and listened to WCCO radio for the weather forecast. I suspect that even when the temperature plummeted to ten below zero, we went to school. Mom kept me well prepared—*all* skin covered. The bottom part of my gear included dreaded brown-ribbed stockings—held up by garters and garter belt—topped by long underwear, flannel-lined blue jeans, and snow pants. The top half of the dressing ordeal: undershirt, flannel shirt, sweater, wool coat, wool scarf, wool hat under fur-lined cap, and several layers of mittens with a leather mitten on the outside to cut the wind. With my feet embalmed in layers of socks, shoes, and (barely) flannel-lined ugly rubber boots, it's a miracle I could walk at all. The flannel lining always wore out quickly, keeping my feet cold.

Dad, like most busy farmers, wouldn't have thought of leaving his chores to give us rides to school. Rides were rare and by chance. Frank occasionally invited us to ride along in his milk truck. Ed's flat-bed truck sometimes came rattling by at Whyte's Corner, transporting Sandy, who had trouble walking—especially in winter. Then everyone happily leaped in, or on, the open back of the truck.

Many mornings, with the mercury hovering at zero, we walked as far as we could in penetrating cold and wind, listening to the howling as snow piled higher along red-slatted snow fences. We measured our progress in distance between electrical poles . . . just . . . one . . . more . . . before the smallest kids began to shiver and cry. When the scratchy wool scarf tied around nose and mouth froze stiff from breathing into it, scraping lips and chin; when shins stung from frozen pantlegs underneath snow pants; when fingers painfully tingled, then went numb, and toes began to throb; we finally gave in. Then we made our way to the nearest house, where someone always invited us, still bundled up, to crowd beside the wood stove until our extremities needled in pain as they thawed slightly from winter's chill. Then we slogged back out into the bitter morning, continuing on our way until finally, at school, we undressed in reverse order: footwear first.

Grabbing a friend, we took turns fiddling with stubborn overshoes buckles and yanking off one another's boots. Pulling overshoes and boots on was one thing; peeling them off, quite another. Whether four buckles or five, fasteners regularly jammed and froze solid with ice and snow. And they broke when forced. But those dreary-dull overshoes were cheap. Some kids had slightly more fashionable, black pull-on boots—no treat to get off, either, often sending the yanker backward into the wall. Later, companies introduced a new substance: plastic boots in bright colors. They proved to be colder, stiffer, and more easily torn than rubber. Then came zip-up boots, some with fur cuffs around the tops. Those metal zippers froze up, too (even when not frozen by the weather) and refused to move in either direction. A pencil then remedied the problem; running ordinary pencil lead over the metal made it move smoothly. Whether buckle-overshoes, pull-on boots, plastic boots, or zip-up boots, we always had a row lined up to dry behind our woodburning stove. Is it any wonder we kids were supremely happy when winter ended?

Along with the outerwear and boots, I longed to remove one more set of items but could not: the brown stockings, whose ugliness was surpassed only by the discomfort of their alternately tight or sagging cling to my skin and the garters that held them up—and gouged me as I sat on my hard wooden chair. No, those, unfortunately, must remain.

To dry our wet and frozen clothing, we draped it over the edges of the round galvanized metal "jacket" that protected the woodburning furnace. It dominated the front of the classroom. We huddled around it for warmth until classes began or longer if our teacher allowed it.

Northern winds buffeted the schoolhouse, rattling the windows. The furnace was no match for the old frame building without insulation, and the fire burned fitfully, leaving the corners of the room frigid. Teaching a well-crafted lesson was one thing; sometimes folks measured a teacher's worth as much by whether or not she could build a good fire—and keep it going. Some had the skill and experience to do both well. But since no fire burned during the night, the school was never very warm. Even so, three times each day we retrieved our winter gear, whether it was dry or not, and hurried outside to play: at morning, noon, and afternoon recesses.

Unbroken snow called for "cutting the pie" (i.e., making wedges inside a circular path for a lively chasing game of Fox and Goose). Reigning as King of the Hill atop the mountain of plowed snow in the schoolyard was an honor worth fighting for, but it often resulted in scrapes, bumps, bruises, and humiliation. Vying competitors face-washed each other with a mitten full of frozen ice crystals—the grains of snow felt like miniature icy daggers—the threat of which was enough to cause many a child to flee and surrender in terror. Snowball fights sometimes ended with an ice ball blow to the head.

We survived the dangerous winter weather (and recesses) by becoming responsible for our own safety, keeping that in mind as we tackled schoolhouse chores: clapping erasers, washing blackboards, carrying water, and filling the wood box. There, too, we learned lessons. A wooden upper opening in the woodshed at back of the school could be propped open with a stick, creating a window. If someone knocked the prop away, the heavy wood slammed and occasionally bashed the unsuspecting head of the student gathering an armload for the furnace. Fearing the worst, we'd rescue the victim and (luckily) find howls and tears the only result.

Having experienced my tongue frozen to a metal sled apparatus, I knew better than to lick the frosty pump handle at the water well. But others had yet to learn it. "Betcha can't put yer tongue on that white handle—dare ya!" The temptation taught a painful lesson, leaving a bit of tongue behind and a darn sore remaining tongue, too (for some, hot chili for the day's lunch was not a pleasant thought).

In deep winter, the school pump froze solid and was useless.

We had to retrieve water from Rob and Inee's place, at the base of Miltimore hill, about a quarter mile up Highway 48. As third or fourth graders, Verna and I were deemed old enough to plod along. We carried empty buckets in hand one way, then sloshed our way back. Water splashed against our snow pants and froze them stiff as boards, chilling our legs. When we finally returned, our pails were half empty. We clumsily lifted and poured the remaining water into the stoneware water cooler—the gray-and-blue-striped Red Wing with the once-broken-now-glued-together lid—where we all drank from the same chilly aluminum dipper.

Fortunately, the Mothers' Club provided a hot lunch program. Lahoma Hanson, and later, Bernice and Grandma Fairman, cooked. As noontime neared, they arrived with blue graniteware pots of steaming soup, chili, or stew, or pans of hot macaroni and cheese, always accompanied by bread and butter and cookies or cake. They served the welcome homemade meal from the door of the small kitchen, while a chosen student handed out kid-sized glass bottles of icy milk, delivered each day from the creamery.

I was eight years old and in third grade when Mrs. Olga Carlson became our teacher (no relation to our neighbor Ole). She was organized and businesslike. She wore her black and silvery gray hair pulled back into a bun, and rhinestones lined the top edges on the frame of her tortoise shell glasses. She dressed in dull rayon dresses—were they always purple?—and her durable brown shoes clicked over the worn pine floorboards as she moved row by row and side to side across the room. She read aloud to us the best literature of the time, more than likely a result of summer courses she took to upgrade her two-year education toward earning her bachelor's degree. Every day after lunch recess, she'd read the latest 1950s Newberry and Caldecott award-winning books to the entire school.

In the depths of one interminable winter, Billy Fairman brought his movie projector, cartoons, and films. The Fairmans had moved from Chicago, seemingly on top of the latest trends. No one else had ever seen a small movie projector. It was a miniature version of a black and white reel-to-reel. Giving in to our cabin fever, and perhaps her own as well, Mrs. Carlson piled us all into the only place that could be completely darkened—the back hall. The make-do unheated "room" contained a few bookshelves, a large battered wood box, and the linoleum-covered counter where we all washed our hands in the same icy water that filled our enamel washbasin. Nearly thirty kids happily jammed into that space, which soon warmed with expectation. Giddy with excitement, we screeched at the ridiculous antics of Charlie Chaplin, Woody Woodpecker, and Mickey Mouse. After we had watched the films and cartoons over and over again, we begged, "Billy! Play 'em backward!" And he did. We collapsed into uncontrollable fits of laughter as the room shook with joy. It seemed those movies went on for two days straight. The normally subdued Mrs. Carlson seemed as happy for the winter reprieve as we were.

As spring finally neared, restlessness hovered in the air. Card games that had been going on all winter became tiresome. Older kids claimed to be playing canasta; younger kids found it too complicated, so we huddled in corners with storybooks. Mrs. Carlson allowed the cards but stressed, "If you see a fine car pull into the driveway and a man in suit and tie, *put the cards away*! It is the county school superintendent. Everyone *must* be on their best behavior until he is gone." He was easily recognized: only he emerged from a shiny car neatly dressed in church clothes. We did behave, and Valley View and Mrs. Carlson evidently passed his rigorous inspection.

In early spring one year, we had another reprieve: the school bell overturned. At the back of the room, a knotted and tattered rope hung, attached to the giant bell in the belfry. Little kids knew the weight of the bell could lift us right off the floor, and older kids knew that pulling the thick rope too hard tipped the bell upside down. When that happened, someone had to climb up and set the bell upright. Some of the older boys, Sam not among them, quickly volunteered. Not long after they came down, clouds of smoke began to roll toward the ceiling. By the time Mrs. Carlson herded us outside, choking and coughing, the room had filled with billowing black smoke. When she abruptly dismissed school, we gleefully raced toward home for a marvelous unplanned extended vacation.

After the plugged chimney had been repaired, the problem was attributed to those three who had *fixed* the bell. Mrs. Carlson ordered the culprits to line up at the front of the classroom and commanded them to bend over. She recounted their shameful deed. Then, with grim determination, and in stony silence, she whacked each behind soundly with a paddle she'd brought especially for the occasion. With gaping mouths and wide eyes, we watched in stunned silence. The boys' behinds in front of us, we never did see the expressions on their faces.

We were happy for any diversion, and especially spring's arrival, when a magical thing happened on our gravel roads. As frost came to the surface, sticky brown mud—like textured brown paint—bubbled in quivering pools just beneath a thin crust. Springing up and down on the spongy ground was better than forbidden mattress-spring jumping at home. We poked curiously at the mud oozing through narrow cracks and coating our overshoes.

At one point in the early 1950s, public service announcements on the radio brought a chilling threat: polio. Infantile paralysis. We knew one boy affected by a useless limb, and we took precautions. Gatherings were discouraged. We did not go swimming at Windfall Lake and we were not allowed to visit Grandma Bessie in the Ladysmith hospital when she recovered from her broken hip. Polio frightened adults and children alike. Who can forget the films shown in school of children encapsulated in belching iron lungs to aid their breathing?

In the spring of 1952, science offered promise with word of the development of the Salk vaccine. In 1954, the nation called for a massive test using school kids. Over the kitchen table, Mom and Dad debated whether Sam and I should be part of the hopeful experiment, albeit with unknown results. One spring Saturday morning, we showed up at Valley View School to get our dose, along with most of our neighborhood friends; some families remained skeptical and stayed away. However, none of the local kids suffered negative consequences from the vaccination.

Usually, just after Easter, when the roads had dried up enough that we didn't have to wear boots, Sonja chose to wear her brand-new patent leather shoes. I have no idea what prompted her celebratory mood that day or why her mom gave her permission to house her feet in her shiny black shoes. She was happily bouncing when she dropped into ankle-deep muck. Her mom's anticipated reaction was enough to send Sonja into fits. No amount of consoling could quiet her uncontrollable sobs for the remaining miles home.

The next day, Sonja wore her oldest pair of battered Oxfords.

A HOUSE IN THE ROCKS

The first house I shared with my friends was in a pile of rocks. Technically, they were large enough to be boulders, but we just said, "Meet ya in the rocks at recess."

They had probably been there since Dad had started at Valley View. Most likely, they'd been rolled aside in the creation of Valley View Road by hand and by horse power when the dirt road had been no more than a wagon rut. Rocks of any size were never in short supply in this glacially gifted country, and workers hadn't been able to easily move these, so there the boulders stayed—in the ditch on the corner of Valley View Road and Highway 48.

One might think schoolchildren playing among hard, unforgiving rocks somewhat of a safety nightmare. Of course, it was. But school had few regulations then, and we learned to fend for ourselves. One day at recess, I fell on the rocks and arrived home with blood oozing through my white furry hat. A wound I didn't even know I had. Accidents happened, and no one supervised our play outside.

Mrs. Carlson dismissed us for recess in an orderly manner, commanding each grade row by row, "First grade, stand. Pass. Dismiss." First through eighth grade. Once released, we bounded through the back door, hitting the ground in full stride. No one hung around to see what Mrs. Carlson did during her free time. She showed up on the playground only for inordinate amounts of blood, broken bones, dangling limbs, or victims who had lain unmoving for a long time.

We had only so many things to do on the Valley View playground, which boasted a teeter-totter and two swings. Mostly we played games: softball, ante-I-over the woodshed, pom-pom pull-away, and Captain, may I? The slightly older girls, who didn't want to get dirty, got credit for the pretend house in the rocks.

Verna and I didn't mind getting dirty; we weren't "houseplants," our term for girls who didn't want to play in the dirt or out in the cold. Like the boys, she and I listened to Milwaukee Braves baseball games on the radio, and what we really wanted to do was play in the daily softball game. But the big boys considered us too little at seven or eight and not good enough for the game on the worn paths of the pasture diamond. Besides, we were girls. So, we often had to content ourselves by playing house with Margie and Sally Fairman, Mary Ellen Bjelland, Sonja, and Shirley.

In the giant rocks, our enchanted home existed in our imaginations, enhanced by a few simple props: firewood from the school woodshed, scrubby trees, pebbles, weeds growing in the ditch, and other organic matter.

"Care to join us? Please step in; yer just in time fer lunch! Would ya like some nice, crunchy gravel cereal? Oh! Yer feelin' a bit tired? Stretch out and relax on our comfy rock bed."

We mended socks with seedpod thimbles, whisked dirt away with willow switch brooms, brought rock-cooked meals on wood slab trays, and ate snail shells on basswood leaf plates. For dessert we baked choke cherry pie in the rough board oven and served it with mud-water coffee in tiny cups made of rolled leaves.

The squeals coming from the rocks proved so inviting they sometimes lured Stevie or Richard or one of the younger boys to join us. But they seemed unable to get into the spirit of our dwelling. When we tried to teach them how to make raspberry rings by tying long strands of Timothy grass around a finger, adorned with a ripe, red raspberry stone, they didn't appreciate the craft. And when we wove classy dandelion stem bracelets, the boys picked up, left, and did not return.

With a gong of the bell, Mrs. Carlson signaled recess had ended and it was time to get back to our books. We dropped our serving trays, aprons, and domestic duties and left fresh cut flowers to wither on the rock table. No setup, no takedown, nothing but imagination required.

HEY, BATTER!

Verna and I had but one ambition when we were eight years old: to be real baseball players. Verna's sister, Jane, five years older, was the best girl player we knew—so good, the boys allowed her to play. Jane was a fast, accurate pitcher who also ran like the wind, hit with force, and could throw in the long fly balls from the out-field in the pasture. She even chattered with authority: "He can't hit! Hey, *hey*, batterbatterbatter!"

We looked on with envy at the beginning of each day's soft-ball game when captains chose teams. The two boys faced each other as someone threw a bat straight up into the air between them. One captain caught it down low. They took turns alternat-ing hand over hand, gripping until one final hand grasped the top and earned its owner the right to choose the first player for his team. They alternated picks until everyone they deemed worthy was chosen. We noticed younger, less skilled kids slink quietly away when they were ignored. Verna and I yearned for the day the captains would choose us, but they never did, so we settled for re-trieving the foul balls that bounced across the fence into the cow pasture or rolled into the muddy turnaround in front of school.

Left to look on from the sidelines at the ball field's rutted base-lines, scrubby pitcher's mound, and muddy home plate, Verna and I would relive the details of last night's Milwaukee Braves game. In the heyday of Warren Spahn, Lew Burdette, Eddie Mathews, Del Crandall, and Hank Aaron, whenever a player hit a homerun, the Braves' radio announcer would yell, "Holy *cow*!"

As we watched the game at recess, we reviewed rules, base-running strategy, banter, and batting technique, hoping someday soon we'd be selected to play in a real game. Or even a game of 500 or work-up, which was played when short of players. Each player gained points by catching or fielding a ball, working their way up to the batter's box until they were put out. But our fondest dream was to someday be a part of the end-of-year ball-game against neighboring Meteor School. The game alternated sites, and the winner gained bragging rights for the year. We were proud of Jane, the only girl on *either* team. Someday, would we fill her shoes? In the meantime, practice was essential.

Playing ball at my house was problematic. Sam didn't really like to play ball, the nearest neighbor kids lived a half-mile away, and farm kids were always busy with chores, anyway—especially in summertime. On the rare occasions in the fall when they could come to play, we were more likely to have a lively game of hide and seek or tag through the corn fields, getting caught by each other and the dry, rattling stalks of corn. But sometimes Sam and friends relented and played ball. Then geography and topography interfered. Trees and a hill surrounded our driveway on two sides, so my only ball could easily be lost in the woods. And a long row of porch windows lined the house. Badger Creek presented a challenge, too.

Everyone had to scramble to intercept a wayward ball before it plunked into flowing water. A missed catch, or any ball hit long or rolling fast, invariably ended up in the drink. If Badger Creek brimmed with water and the ball landed in the main part of the stream, our only ball would be whisked through the long culvert and into Jerry's sheep pasture. That presented real trouble—long rows of beehives and, sometimes, rams. Now and then our ball disappeared forever when it tumbled into the foamy flow. Even if retrieved, a leather ball had a shortened life span after being dunked and water-soaked.

I found a solution by practicing ball at home by myself, throwing an ordinary five-and-dime store sponge-rubber ball against our back porch. The porch had irregular log sides, making fielding tricky when the hard ball bounced back unpredictably—and quickly. I learned to scoop up grounders, field pop-ups, and snag line drives by throwing just so. I became proficient at hurling and then racing into position, usually catching the thirty-nine-cent ball before I had to sprint toward the creek to save it. Mom and Dad endured the constant thud of the ball as they sat in the kitchen sipping coffee when after a meal I rushed outside for more practice. They never complained. Nor did they mention the mud blots left on the freshly stained brown wood. My endless practice paid off one day when I chased down a foul ball at school. Right on target, I fired the ball back to Bobby. He couldn't help but notice, commenting, "Hey, she's got a good wing!"

Only owning a baseball glove or being allowed to play would have made me happier. I longed for a real baseball mitt. The best I could do was rummage through the box of Dad's winter leather gear; there I found a big work mitt. It was nothing like a real glove, but it did make that smacking sound I wanted to hear when the ball hit leather. I would have died if my friends saw me wearing it, so I used it only when alone. I drooled over the gloves pictured in the mail order catalogs: Spalding, Wilson, and Ted Williams brands. Even the least expensive glove cost more money than I had. All I could do was save twenty cents of my allowance each week, after keeping part of it for candy, of course. Summer would be well over before I had enough money, and Christmas and my birthday were too far off. Finally, I grew tired of the long wait saving for a glove and sent two dollars for a Louisville Slugger bat, only to find it was not sturdy like I'd imagined. We used it anyway—until it broke.

Ball, mitt, bat—none authentic, all disappointing. But one year, Mom bought some high-top canvas shoes—black, of course. I did not care that they were boys' shoes; in fact, I was thrilled. They proved tough enough for creek wading, rock leaping, mud-pie making, getting the cows, stepping carefully through the barnyard, tearing around in the woods, and, especially, playing ball alone against the back-porch wall, imagining myself a baseball hero. Perfect shoes.

Verna and I figured by the time we were in fifth grade we'd be good enough to play ball at recess and maybe even for the game with Meteor School. Alas, by then we had to attend school in town. To our surprise, the girls in fifth grade in Exeland dressed like . . . girls! They wore dresses and pedal pushers and played with hula-hoops. Town girls didn't play ball. Verna and I realized we wouldn't be able to play softball at school after all. But we stubbornly agreed, "cross our hearts and hope to die," we would *never* carry a purse!

HALLOWEEN MASQUERADE PARTY

Verna, Richard, Margie, and I plotted what we might do for Halloween, especially after Richard showed us the remains of a bar of soap and a round disc of paraffin he'd taken from his mom's jelly jar that morning (paraffin was harder to remove from glass than soap was). I opted for the lesser evil: "We could soap Mrs. Carlson's car windows," this from my bold, brave (at the moment) voice.

"Yah," said Verna. "But what if she found out it was us? Besides, she's not really that bad."

"Or we could rub soap all over some bachelor's windows Halloween night?" Margie offered, tentatively. But none of us could think of anyone we knew who really deserved such treatment. Richard stuffed the soap and wax bits back into his pocket for further contemplation.

Some of the older kids told tales of tipping over outdoor toilets when occupied, or worse, moving the entire outhouse away from the hole. In the darkness, the victim would fall into horrid emptiness—until hitting the bottom. We'd also heard stories of farmers who sat with a buckshot-loaded gun across their knees, in wait for pranksters.

Since none of us seemed to display the requisite bravery, we did nothing but carve pumpkins and make wooden-spool noisemakers, popcorn balls, and pans of chocolate fudge at home. Once Sam and I branched out—pulling our own taffy. None of us went trick-or-treating. Thus, most of our Halloween entertainment was provided by the masquerade party at Valley View schoolhouse and the atmosphere created by Mrs. Carlson as Halloween neared.

Every day, Mrs. Carlson read stories to set the mood, and my imagination ran wild as I pressed my waxy crayons harder still into limp manila paper. To the words of Edgar Allan Poe ("The Raven," "The Fall of the House of Usher," "The Tell-tale Heart"), I created the forms of snarling black cats atop fence posts. While she read James Whitcomb Riley's poem about the "little orphant" Annie's ghosts and goblins, I drew shadows of hoot owls on skeleton-like trees. Accompanied by Washington Irving's "The Legend of Sleepy Hollow," my fat orange crayon drew a fat, full moon and sneering scarecrows. This story was my favorite, as it featured a two-horse chase, one ridden by the schoolteacher— with the outlandish name of Ichabod Crane—after he'd sought the hand of the equally oddly named Katrina Van Tassel.

Then we decorated the windows and walls at the schoolhouse with paper jack-o'-lanterns and our pictures, and, finally, the day itself came.

After the evening milking, our car approached the shining schoolhouse windows in the darkness of late October. The paper jack-o'-lanterns grinned toothily from each window, looking much more fearsome than they had in the daylight, while flickering carved pumpkins stood on each side of the front and back entryways. Even the way to the outhouses was illuminated by a leering orange glow along each path. Those pumpkins had been surprises for the night, set up by the Mothers' Club during their PTA meeting before the party started.

The evening began with games, mostly food-related. Bobbing for apples was a bit too messy, not to mention unsanitary. Instead, and not much more sanitary, we hung apples on strings and competed—who could take a bite without using their hands?

Next, after eating a soda cracker, who could eke out the first whistle? Lined up with our hands clasped behind our backs, we passed an orange from under one chin to the next, in boy/girl/boy/girl fashion. Smaller kids, blindfolded, tried to pin a gaping grin onto the face of a pumpkin. The blow-up-the-balloon contest was won when the first contestant filled a new balloon with so much air that it was the first to pop.

Mrs. Carlson gave small rewards, if only being openly proclaimed "Winner!" (Kids in our neck of the woods in the 1950s weren't demanding.) As with the games, Mrs. Carlson would be judge of the carved jack-o'-lanterns we'd brought from home and our costumes.

Kids excitedly adjusted last-minute costumes and masks. Some had made a mask at home by attaching string after cutting out the shape printed on the back of cereal boxes, just in time for Halloween. Mrs. Carlson lined up kids, revealing an abundance of cowboys and cowgirls, Indian chiefs, braves, and princesses, a clown, and some ghosts. Baseball players and sailors, circus fat men, and a mop-wielding washerwoman . . . or was that Little Orphan Annie herself? Mary Ellen, who was annually terrified of Santa, had dressed as a scary witch with a tall, pointy black hat and broom. She had perfected her high-pitched screech so well that she would reuse her costume each year.

Over the years, I'd been a raggedy-patched hobo, with a homemade corncob pipe and a bundle tied in a red bandana attached to a stick I carried over one shoulder; a farmer wearing a straw hat, flannel shirt, and bib overalls; and Davy Crockett. That year, I had my own Davy Crockett belt and buckle, and I longed for a real coonskin-tailed hat, so popular at the time. But I had to make do without—though the hatchet-axe I carried was real. Never was I a fairy queen or a princess.

Kids laughed uproariously when Grandma and Grandpa

Fairman showed up, she in flannel pajamas, he in nightshirt and night cap. And Lahoma and Ole Hanson arrived dressed as the foot-stompin' hillbillies we'd seen in Ma and Pa Kettle movies.

With the games and judging over, it was time to eat. The Mothers' Club moms piled the table with blue-speckled roasters of macaroni and cheese, hamburger-macaroni-tomato goulash, crocks of brown baked beans, and heaped dishes of creamy scalloped potatoes; tuna, salmon, cheese, and bologna sandwiches; red Jell-O with bananas and whipped cream on top; cookies, candy corn, frosted orange cupcakes with a spooky black cat on each, and popcorn balls made from scratch with home-cooked sticky goo. For drink, the moms poured what I'd come to call the "Valley View Schoolhouse Punch"—for this occasion only, in orange.

THE GOLDEN BOOK OF FAVORITE SONGS

Besides the songs of Stephen Foster included in our yellow *The Golden Book of Favorite Songs*, a book millions of school kids used in that era, we sang tunes capturing the moment and mood of the times. Younger kids, like me, embraced the novelty of song and looked forward to the change of pace singing provided, even though we could not yet read the words in the book. When I was in first grade, I shared a school bench for singing with an eighth-grade girl, each of us clutching half of the book during the first activity of our Friday mornings.

I would like to say that with smiling faces we all looked forward to those stirring sessions of song. That our voices burst forth and filled the room with melody, ringing out through Valley View's east windows. Or even that a single voice, clear as birdsong and sweet as a nightingale, led the muddle of music-less farm kids. Or that from that schoolhouse came a musician of exquisite talent. But none of this was true. We tried, but our chorus sagged, as limp and listless as the drooping rope attached to the school bell at the back of the room.

Mrs. Carlson did not play, or did not want to play, the black piano that stood at the front of our schoolroom, so she delegated the task to Mrs. Jacobson. Mid—short for Millie—walked across the cow pasture that separated our schoolhouse from her farmhouse. She took her place on the black bench, with her back to us, and swayed at the tall piano as she pounded out songs she knew so well that she could gaze out the single window looking toward her barn and farmhouse.

Older boys intentionally wore dull expressions to show their disdain for the activity so beneath their dignity, hoping only that it would soon be over. They did not look forward to the math lessons that would follow with any great enthusiasm, either, but this was the worst. Some boys mouthed the words silently in an attitude of cooperation, while thinking only of their BB gun or the pocketknife they carried. Most kids came to life only at the conclusion of singing when all ages were allowed to boisterously belt out a lively "Bill Bailey" or "I've Been Working on the Railroad." *Everyone* loudly sang the upbeat songs with a rollicking rhythm.

The Golden Book of Favorite Songs contained rounds, nursery rhymes, and patriotic songs. There were hymns, "Battle Hymn of the Republic" and "Come Thou Almighty King," as well as the old favorites "Oh My Darling, Clementine" and "Grandfather's Clock." We, Yankees from northern farms who had never crossed the Mason-Dixon line or even heard of it, sang songs of the South, such as "My Old Kentucky Home." We knew little, if anything, of human bondage, but we sang such songs as "Hard Times (Come Again No More)." With little background in history and no experiences of war, we sang of Civil War battles, glorifying the Confederacy with "Dixie," "John Brown's Body," and "Carry Me Back to Old Virginny." (But where was the puzzle piece called "Virginny" on my cardboard map of the US?)

Perhaps our favorite songs of all were for the Christmas program. We reveled in the increasing excitement as it drew near, practicing "Jolly Old Saint Nicholas," "Jingle Bells," "Rudolph, the Red-Nosed Reindeer," and "Frosty, the Snowman."

One December morning, Mid thumped on the piano, hard, trying to maintain our focus one last time. The Christmas program was fast approaching; we'd grown a bit weary of rehearsal. We watched her hefty frame move side to side on the creaky bench, mesmerized by her behind swaying in rhythm. As she broke into a rousing chorus of "Joy to the World," our attention turned to a resounding crack. All four bench legs suddenly folded inward in a stunning display of symmetry, dumping Mrs. Mid Jacobson's rear-end onto the unforgiving schoolroom floor.

There was a collective gasp. A split second of dead silence. And then every child broke into uncontrollable laughter. Some fell from their desks howling. Some could not stop, as Mrs. Carlson commanded while hurrying to Mrs. Jacobson's aid. Fortunately, she was unhurt. Laughter reverberated even as Mid rose to her feet and gathered her coat and a morsel of dignity; even as she rushed red-faced from the schoolhouse without a word. Laughter still ringing in her ears, Mid marched home across the pasture.

Mrs. Carlson gave us the tongue-lashing we deserved and would not soon forget. We hung our heads in shame, at the same time fearful we'd burst suddenly into yet another fit of the giggles.

In short form, the piano bench was sturdily repaired—just in time for the Christmas program. On that night, Mrs. Jacobson returned, holding her head high. Without flinching, she slid into place on the restored bench. With heroic effort all around, Mid displayed perfect composure, we contained our snickers, and the Christmas program progressed and concluded without a hitch. The curtains closed while the schoolhouse echoed with the applause of an approving audience for the resounding finale: "Joy to the World."

THE CHRISTMAS PROGRAM

For as long as I can remember, every December, Mid Jacobson's backside swayed side to side accompanying a chorus of kids. Following Sam, five years my senior, meant I'd been a seasoned Christmas program veteran even before I reached school age.

On one particular Christmas, in 1951, the schoolhouse hummed, crowded with the excited voices of young and old. Dad stood with the other men at the back of the room, and Mom jammed uncomfortably on the bench of a desk too small, near the front of the room. The usual lighting, provided by round globe lights hanging from the ceiling on long chains, was partly hidden by a festive false ceiling. Long red and green strips of twisted crepe paper hung from each side of the wall leading to the center; an enormous, red foldout tissue-paper bell hung in the middle. The Valley View community did not waste money, so the faded decorations had been used for years, crepe paper always rolled up carefully post-use and stored in cardboard boxes in the school-house attic.

As a four-year-old, I sat nervously beside Mom. My nose took note of the deliciously odd mixture of odors in the room. Traces of brewing coffee and pipe tobacco radiated from Rob's coat, mingling with the tempting scents of chocolate frosting, bologna sandwiches, baked beans, and sour dill pickles. And the unmistakably sour odor of cow manure. Someone in the room still wore his barn boots.

School programs began at eight o'clock, after farmers had milked their cows. Precisely on the hour, the bulky figure of Miss Stern, the teacher at the time, appeared on stage. With a stern look and a wave of her arm, she brought the room to obedient silence. "Our Christmas program will begin in just a moment," she boomed.

Then no-nonsense Miss Stern strode down the narrow aisle

between the straight rows of stationary desks. She was coming for me. Taking my hand and nodding to Mom, she marched forward with me in tow, barely missing Grandpa Fairman's outstretched cane as we neared the front to face the audience. With her free hand she lifted a wooden pint-size chair with the ease of a Sumo wrestler and swung it into place. Her wide face tilted down at me; her heavy underarms flapped as she lifted me to stand on the chair. Behind me, rough burlap curtains put in place for the program hung stiffly on a wire strung across the front of the room. I heard tittering laughter and nervous whispers rising from behind the curtain—twenty-nine kids huddled there, including Sam, waiting for their appearance on the makeshift stage. How I longed at that moment for next year, when I'd be behind the burlap curtain with the other kids, not on stage alone to fulfill the custom of someone's little brother or sister reciting a piece. Then Miss Stern turned abruptly, the flowing skirt of her flared dress swished wildly and created a breezy current. In its wake, I was sure I would be swept from my perch. When I found myself still standing on the chair, I knew it was time to deliver the lines I'd practiced over and over:

I'm not so very big,
Not so very tall.
Welcome to our program,
Welcome one and all.

With a burst of applause launching me forward, I scurried bashfully toward Mom, who beamed with approval. Thus began the 1951 Christmas program at Valley View School. At its conclusion, everyone had to agree: even if Miss Stern had her shortcomings, she put on a good show!

A few years later, in anticipation of the upcoming Christmas program and holiday, I took advantage of a rare snooping opportunity when Mom and Sam helped Dad with outside chores. Mom usually stayed in the kitchen cleaning up after supper, so I knew she wouldn't be gone long. I had to act quickly. Standing on a chair, I reached high into the only closet in the house, located in Mom and Dad's bedroom. It didn't take long before I uncovered the Sears shoebox on a shelf above my head. Hoping it contained anything but shoes, I opened it and discovered something especially shocking. Yes, shoes. But it was worse: a pair of black, velvety Mary Janes! Not a baseball glove, or a cowboy shirt, or a game, but girlie shoes. Horrified, I rooted around further, hoping to uncover something else. I did.

Doggone it! A bright red corduroy jumper, a matching hand-sewn *purse*, embroidered with bright flowers all around, and a frilly, white blouse with a rounded lacy collar—a *Peter Pan* collar, something Mom was forever trying to slip into my wardrobe. Mom had obviously spent a long time working on these gifts. She had sewn all of it, probably during hunting season. And with

The 1954 Christmas Program at Valley View. Peggy is fifth from right, wearing the white shirt and plaid skirt.

a few Christmas programs under my belt, I knew she meant to give me the outfit before that year's show, less than a week away.

I suspected there must be better presents somewhere, but the fun of searching had vanished. Which was worse? My disappointment, the guilt of snooping, or disappointing Mom? What a dilemma. On one hand, I cared deeply about avoiding ladylike clothes; on the other, I was not heartless. And I had to wear something to the Christmas program. Those clothes. I guess. But those *shoes*?

"I found the shoes!" I blurted, upon Mom's entrance into the kitchen. "And I'm not wearin' those shoes to the Christmas program! Or anywhere else *either*!" I added for emphasis.

Mom listened, a look of disappointment crossing her face. Remorse slashed through me, and I felt even worse when she didn't even debate. That evening, she silently wrapped the shoebox in brown paper for return in the next day's mail. I had no heart for further protest. I swallowed hard, keeping the extent of my snooping a secret. A few days later, when Mom unveiled the outfit for the next day's big night, I responded with as much surprise and happiness as I could muster. If Mom knew *I* knew, she did not say so.

At the Christmas program, I stood behind the burlap curtains with the other kids, wearing my red corduroy jumper and frilly white blouse with the lacy Peter Pan collar. When I arrived, Mrs. Carlson said, "Well, Peggy, don't *you* look nice!"

After Mrs. Carlson's compliment, I tried to glare at Mom. But when I saw Mom's face beaming, I knew this was a battle I had lost: for once, her daughter had received Mrs. Carlson's stamp of approval for appearance.

As we stood waiting behind the curtains, I noticed some of the girls, like Shirley, were wearing white ribbed stockings for the occasion, unlike our everyday brown ones. Not that I thought changing the color would help—I hated long stockings, no matter what color.

The curtains parted, drawn by grinning boys with slicked-down hair and stiff new shirts. Three rows of kids stood ramrod straight and still, except for their moving eyes that searched for familiar faces. I spotted Mom; she clutched my matching purse with the embroidered flowers.

On the piano bench, Mrs. Jacobson's red dress moved in rhythm to her fingers thumping out notes on the upright piano. Our cheery voices sang, "Up on the Housetop" and "Santa Claus is Comin' to Town." On stage in the front row, little Mary Ellen began nervously twisting the skirt of her plaid dress, rolling it higher and higher around her hands until she exposed first her garters, then the garter belt holding up her white-ribbed stockings. Snickers escaped from the audience. Grandpas, grandmas, moms, dads, and those with no children at all smiled and laughed through the recitations, plays, and singing, with little regard for quality. Even Ole and other bachelors in the community snorted loudly when something tickled them.

When the program ended, Santa stomped through the back door; we knew it was really Ralph Woods, but we didn't care. His thunderous "ho-ho-hos" scared Mary Ellen, and she began to howl. Just like last year, she ran screaming to hide under the table at the back of the room. Santa's pack, which looked remarkably like a pillowcase, held small items wrapped for the school kids' gift exchange, most purchased in Exeland. We easily guessed the contents of each package by its shape. Cylindrical packages meant either a game of pick-up-sticks or a rolled-up coloring book. Long, narrow boxes contained a small pencil box, a set of checkers, or black wooden dominoes. Other packages wrapped up crisp lined tablets with pictures of Roy Rogers and Dale Evans on the front or yellow and green boxes of colored pencils and crayons.

After Santa had delivered the wrapped parcels, he dug into his fat sack to retrieve brown paper sacks with tightly twisted tops, handing one to each student. We wasted no time peering inside; we knew they were always the same. Each promising bag, round and full, invariably contained lots of peanuts in the shell, a few colorful pillows of hard candy, a piece or two of chunky peanut brittle, and two or three chocolate drops. But we were happy anyway, amused as Santa happily presented extra bags of candy to some of the grandpas and grandmas in the crowd.

Then the eating began! The men moved tables onto what had been the stage at the front of the schoolroom and the Mothers' Club moms set up the usual potluck fare, with the ever-important addition of cakes slathered with chocolate frosting and, my favorite, cakes covered in white gooey-yet-fluffy seven-minute frosting, made from beaten egg whites. Scandinavians in the neighborhood brought festively dainty open-faced sandwiches: slices of dark rye bread spread with cheese from jars of Cheese Whiz and topped with sliced green olives and pimientos.

All too soon, one by one, the men went outside hoping their cars would turn over and heat up. Dad stepped out the schoolhouse door into the frigid night to join the pack. Moms began to retrieve small children who'd been put to sleep atop a huge pile of winter coats at the back of the room. Families bundled up, pulled on boots, and trudged toward their loudly thumping cars, overshoes crunching in the snow. When Dad returned, blowing steamy breath from the cold, we were ready, stuffed into lumpy layers. Then we stepped into the starry night. Our Scandinavian neighbors were much too reserved to think of shouting out a hearty good night or wishing us Merry Christmas, but we understood their brief waves just the same.

Car doors slammed, bare tires spun on the icy schoolyard, and lines of car headlights crept off in all four directions. After we all piled into our '49 Ford, Dad cranked the car around slowly, and the snow screeched sharply as he accelerated on blocklike tires. Our car bumped south on Valley View Road, crossed white and frozen Badger Creek, then stopped to drop off Grandpa Sam and Grandma Bessie. We drove on past Jerry, Millie, and Shirley's house, all the while reliving the best moments of the night.

My mind joyously reviewed my unsurprising gifts: pick-up sticks and the just-as-supposed candy. Santa's arrival had been predictable. The recitations had been familiar yet entertaining. The music and singing had been no better than usual. The tree had been scrawny and the decorations faded, handmade, and simple. No matter. Every imperfect moment had been absolutely perfect.

We rounded the bend at the end of our driveway and crossed our bridge over Badger Creek. The car lights shone into the windows of our home. Smoke was rising from our chimney. I knew after Dad stoked the fire that our house would soon be cozy. Stepping from the car into the frosty night, I noticed the twinkle of a million stars and light glistening on the sparkling snow. Had it ever been so bright?

Then Dad pulled the back-porch door tight behind us as the intoxicating fragrance of our fresh-cut Christmas tree greeted us. I crawled into the warmth of flannel sheets on my bed, pulling Grandma Bessie's patchwork crazy quilt up to my chin as my cat snuggled next to me. With jingle bells still ringing in my head and the taste of peanut brittle on my tongue, I dreamed of the two-week Christmas vacation just begun and the holiday itself, just a few days away.

VALENTINE'S DAY

One February morning in 1956 dawned as usual—dark and cold. Leaping out of bed and rushing behind the toasty wood stove, I pulled on warm corduroy pants and, for today, my bright red sweater. After Mom's breakfast of hot oatmeal, Sam and I donned our outdoor gear. We waved goodbye in the dusky dawn of day at seven o'clock, carrying thick, brown envelopes. It would take us about ninety minutes, including play, to get to school.

After a ten-minute walk, at our mailbox we greeted Sonja and first-grader Johnny, my sled-shoving replacement. Today the four of us bounced along with more excitement than usual, taking turns towing our sled up the first short hill. We needed only the one; we'd soon be sailing down the long Carlson hill just ahead. At the top of the hill, in the still-pale light, we looked through haze toward Exeland five miles away. As the sky lightened, the sun broke through. Clear and cold, this was going to be a perfect Valentine's Day!

With no patches of gravel on the road to slow the metal runners, it looked like we would coast all the way down the half-mile hill. We threw our brown-paper envelopes stuffed with valentines aboard the sled. As the oldest, Sam piled on first, belly down, gripping the steering handles; then came Sonja, then me. Little Johnny sent the sled on its way with a mighty push, then took a flying leap aboard before the sled gained speed and dipped down over the crest of the hill. Johnny grasped at our coats, hanging on for dear life. It was an exhilarating glide to the bottom, but the sled had no more than slowed when Johnny's loud wails pierced the air.

"Stop yer bawlin', Johnny," Sonja scolded, as we tumbled off. "What's wrong with ya?"

"Oh! Oh, no-o-o!" Johnny blubbered miserably, pointing his shaking mitten back up the hill.

One look explained Johnny's woes. In the distance, strewn between the banks of snow lining the road, lay a crumpled brown envelope and tiny bits of pink and red. We'd forgotten to teach Johnny how we safely stowed our packets under Sam's belly. The four of us tramped back up the hill, collecting the damp missives until Johnny was satisfied we had salvaged all of his precious cargo.

A Valentine's Day disaster of such proportions was not to be taken lightly.

I had dreamed of selecting valentines for weeks. Should I choose the ordinary twenty-nine-cent packages at one of the local grocery stores, as most did? Or should I order from the brightly colored pages in the mail-order catalogs? Each year, I pored over the assortments available, begging Mom to let me buy the more costly valentines that had a red heart-shaped sucker tucked in. I longed for that special package but usually settled for an economy kit from Sears or a big valentines book sold at the five and dime in Ladysmith. I would cut individual valentines from the book, fold them, then slide each into an envelope.

But first I spread the thirty valentines across the kitchen table and deliberated carefully over each. With my list of school names in hand, I chose just the right one for each person in every grade, one to eight, and a slightly larger one for Mrs. Carlson. Silly sayings, ridiculous riddles, witty jokes, and various themes—from puppies to ponies, posies to kittens, cowboys to cars. With a stumpy crayon or pencil, I marked each valentine with a "To" and "From," licking each envelope and sealing it tight for someone special. Sam, Sonja, and Johnny had done the same, Johnny's mom helping him with the writing.

Back at the bottom of Carlson's hill, we clutched our packets more tightly and continued our hike down the road, sled trailing behind us. Picking up seven more comrades, so it went: spirits high, our happy band hurried along toward school, eager to deposit our valentines in the Great Big Valentines Box.

Dazzling valentine cutouts lined the sunny east windows and greeted our arrival. Twisted crepe paper streamers of pink and red lined the ceiling. On a table at the front of the room stood the gorgeous valentine box. This year Mrs. Carlson chose eighth-grader lucky-ducks Jane and Sandy to decorate the room and the box. For weeks, it seemed, they had cut and crumpled paper into hearts, flowers, and cupids. The smell of peppermint still wafted through the classroom from the white paste they used to affix each piece, magically transforming the box from brown cardboard into vivid pink, red, and white. We proudly emptied our valentines into it and waited impatiently until afternoon when Jane and Sandy finally tore it open. Then we became postmen, gleefully delivering mail to our classmates until the box was empty.

Finally, at our wooden desks, Mrs. Carlson allowed us to open our own envelopes. Our eyes scanned about the room to see who got what. Did my friends like the valentines I'd chosen for them? Who had the biggest valentine? *Did Sonja just open a valentine with a sucker in it?* What special valentine would I receive? This one bulges; is it a candy heart? What does it say? "I'm yours!" And this one? "Be mine?" *Who dared* to give me the candy heart that said "Kiss Me!"?

We ended our day munching heart-shaped sugar cookies frosted in pink while gathering our treasured collections in the sugar-sweet room echoing with laughter.

SMOKEY BEAR

Like many rural schools, Valley View had been aptly named in appreciation of its location in the natural world, where the Blue Hills softened and sloped eastward to the Chippewa River Valley. Other local country schools identified themselves with similar affection: Meadowbrook, Windfall Lake, Wooddale, North Star, Glendale, Lone Pine, and Meteor—named for a fallen meteorite.

"Meteor" as a moniker featured often in our neck of the woods: Town of Meteor, Meteor Hill, and Meteor Fire Tower. The fire tower, built in 1934, rose on galvanized steel girders one hundred feet in the air, overlooking the vast stands of hardwood it protected. A fire "spotter" climbed the one hundred-foot ladder atop the tower, straight up to a seven-by-seven-foot room. Windows in all directions allowed his keen eyes to scan the forest expanse in springtime, as leaves appeared, and in fall, when dry and windy conditions could mean trouble. Rising smoke, a lightning strike, or a fire burning in a location not noted by the purchase of a burning permit caused alarm.

Dad sometimes drove to the tower for a visit. Although spotters had to be attentive to their task at all times and were not allowed reading material or a radio, except for fire communication, they briefly welcomed visitors. Emil Oeleke and his son, Willard, who followed him on the revered job, drove from Exeland to man the tower in season, often stopping by Rob and Inee's place afterward. Dad sometimes joined them for the tale-telling on fire tower days. The spotters held precious memories of stunning views from above that paradise of pale-green springs and red, orange, and yellow autumns. They cherished the time spent alone watching wildlife: deer, bear, coveys of partridge, and flocks of birds migrating at eye level.

Between the Meteor Fire Tower and Valley View School, one of the many hills rose above the others, offering sweeping views to the north and east; local folks called it Bobcat Hill. (Every teen with a car on a Saturday night also knew the spot as a romantic stop to watch the stars, maybe even a meteorite shower.) Our community proudly marked that hill with a sign noting it as one of the highest points in Wisconsin: elevation 1,801 feet. Bobcat Hill held its own with the state's highest point—measuring only 151 feet lower than Timm's Hill, located in the county bordering to the east. Bobcat Hill was forested with hardwoods, as was the rest of our area, and, as the crow flies, Valley View School was just a mile away.

Mrs. Carlson invited an old-timer lumberjack, Hank

Hendricksen, to demonstrate the importance of our forests in our local logging history and its impact on our economy. Hank had logged white pine during the cutover heyday. He'd also been a "river hog," wearing his high-water pants while driving logs on the Chippewa River to float down to the mill in Chippewa Falls. Hank held our rapt attention while displaying curiously named tools from the logging era: cant hooks, peavey poles, and a timber cruising stick, which he used to measure the value of standing timber. With one of the boys, he demonstrated the two-man crosscut saws. I coveted Hank's hobnail boots. Sharp spikes protruded from the soles, to gouge into spinning logs. Oh, to have those boots to leap across logs entangled in a log drive!

To instruct us on the importance of fire prevention, one day Mrs. Carlson invited a forest ranger. He arrived in his spiffy forest green Malone-wool uniform. Dad had already schooled Sam and me on the danger of fires, especially forest fires, and Mrs. Carlson and the ranger reminded us that fires could destroy our rich timberlands. The ranger demonstrated the hazards of fire by putting paper into a Pyrex dish, telling us fire needed three things: fuel, air, and a spark. When he dropped a lighted match onto the paper, the fire blazed into a flaming fury, soberly reinforcing how quickly a devastating fire could spread. With fanfare, he clapped a lid on tight. Air cut off, the fire snuffed out. Magic.

While Mrs. Carlson taught us the words to the "Smokey the Bear" song, distracting us, the ranger disappeared momentarily into the back hall. As we sang, Smokey miraculously appeared! His legs were clad in green ranger pants, his furry chest was bare, and a wide-brimmed ranger hat sat on his head. Mrs. Carlson handed out Smokey Bear comic books, song sheets, an official Smokey Bear wooden ruler and bookmark, and a badge proclaiming each of us an Official Junior Smokey Ranger. Then Smokey marched around and led us in a rousing chant: "Only *you* can prevent forest fires!"

That day, I walked home clutching my Smokey Bear paraphernalia to my heart, now possessing an undying fondness for this bear, and loudly singing the chorus of his song:

Smokey the Bear, Smokey the Bear,
Prowlin' and a growlin' and a sniffin' the air.
He can find a fire before it starts to flame!
That's why they call him Smokey,
That was how he got his name!

THE SCHOOL YEAR COMES TO AN END

On our walks to school in April, we would scan the roadside for fuzzy harbingers of spring, scouting the location of willow bushes. When we found swelling buds with their roots soaking wet beneath the snow, we ventured onto thin crusts of ice over the watery beds. We didn't care if we fell through and drenched our feet while twisting off the limber stems. Clutching our prized bouquets of soft, smooth nubs, we burst through kitchen doors shouting, "Ma! Look! Pussy willows!"

As spring settled in, we left for school in the full light of day. Mornings dawned clear and fresh, sunlight streamed through woodland cobwebs, and branches sparkled with dew. Now we easily caught the slow-moving toads, still sluggish after winter, that warmed themselves in rays of morning sunshine. Robins yanked brown worms from the soft wet sod on plowed fields. We walked to school listening to the call of disorderly crows, croaking frogs, and the high-pitched din of spring peepers as we peered into the trees for the first sight of anything green, knowing it was sure to be the lilylike leaves of leeks—wild onions. Leafy clusters had already thrust through matted dead leaves. I hated the taste of onions but celebrated the lowly leek for its early appearance.

At school, when the garlicky breath of a boy anchored in the straight row seat behind us drifted into range, we knew he had been tempted to taste the leaves. Squirming in our bench seats, we heard one voice hiss angrily, "Shut yer trap and stop breathin' on me!" All the while, we hoped he had given himself a churning stomachache for his thoughtless indiscretion.

"Keep your eyes peeled for mayflowers," Mom suggested one morning as we left for school. "I expect they'll be out any day now."

We didn't need reminding. Hepatica—mayflowers—were one of the first woodland flowers, never far behind the leeks. We often found them next to the trunk of an awakening maple tree, as if huddled there for warmth. Tri-lobed and liver-shaped, the purple- and green-blotched leaves stayed green all winter, tucked under a thick layer of leaves. Overnight, clusters of tiny pink, lavender, and white flowers poked through in flawlessly formed bouquets.

The timing was perfect. On May Day, Mrs. Carlson helped us cut and fold pieces of stiff construction paper into cone-shaped baskets. Then we attached paper handles by smearing on the gooey peppermint-scented paste and scrawled "Happy May Day" messages on the sides. At recess, we crossed Valley View Road into Thorson's woods, squealing with delight at the first bloodroots thrusting their white petals skyward. But we did not pick them to fill our baskets; they would shed blood-red liquid from their stems. Besides, we knew this flower would lose its petals quickly once taken from its woodland home. Instead, we chose the velvety petals of yellow and blue-purple violets, and yes, the first mayflowers, too.

On the way home, Shirley and I crept silently toward Grandma Bessie and Grandpa Sam's front door, hiding our baskets behind us. Then we ran up the steps and across the wide porch to hang our gifts on the doorknob, quickly knocked, and dashed away before anyone could catch us for a kiss.

In late spring, once the first bike bravely appeared at school, *everyone* rolled bikes from storage where they'd spent the winter at the back of the shed. Kids oiled their own gritty bike chains, patched flattened inner tubes, and pumped up tires with air. We tested our coaster brakes on the first long hill, celebrating winter's release after shedding our bulky gear, lightened and gleefully free.

Sam and other boys turned their bikes upside down in the schoolyard to fiddle with exposed mechanical parts. Some bikes had chain guards, but Sam had "stripped down" his bike for speed—and rolled up his pant legs, keeping them from wrapping up into the rotating chain. It didn't always work; boys freed their pants by ripping them out of the chain, resulting in torn pants and black greasy stains. Or worse: when pants caught in the chain of a bike on the move, it caused a sudden halt and threw the rider over the handlebars headfirst for a nasty tumble on gravel roads.

In third grade, I eagerly rode my own bicycle—even if it was an ancient, heavy bike Dad had bought for economy over looks when he'd found an ad in the *Ladysmith News*. Riding made the trip to school much faster—when there wasn't mud, that is. Avoiding ruts cut deep in the mud-caked road, we cruised toward school by steering to the ridges on high ground, avoiding patches of gravel where we could easily spin out and crash. After a fast, airy ride down steep Prilaman hill, we coasted effortlessly over the long, gently sloping grade, then cranked hard on the flats toward school. At Badger Creek we braked sharply, skidded to a sideways stop, propped our bikes on kickstands, and hurried to the bridge.

A high flow of iron-brown water rushed under the bridge. It tumbled over piles of rocks into placid pools. We eyed the dark basins, hoping to spot a brook trout before our looming shadows sent fish skittering for cover under the overhanging bank edge. From the roadside we peered into the swampy woods, hoping to be the first to retrieve brilliant yellow marsh marigolds, knowing the bouquet would be worth the cost of drenched boots and soaking pant legs. Could snowy white trilliums and show-off jack-in-the-pulpits be far behind?

Once the ground dried out, softball games resumed from the months-long hiatus. The scent of freshly oiled baseball gloves permeated the air. A prized wooden bat, cracked by Bobby's home run last year and mended with a layer of tape, might last one more season before splitting apart. And it was always a good idea to add more tape to the handle for a better grip. But sometimes kids required more than tape to repair their worn equipment, and a trip to Rob's workshop saved the day for many: "D'ya think Rob could stitch the torn ball?" "Yah—and fix the rips on this webbin'?" "An' the laces on my glove?" Yes, he could. And he did.

"Batter up!" Richard yelled, as fielders dashed to the outfield to save precious recess time for play. We needed to sharpen our skills for the once-a-year-game with Meteor School only a couple of weeks away. The practice brought results too—our pitcher struck out many of the shocked Meteor boys. They weren't used to hitting against a girl. Hooray for Jane! Valley View 15—Meteor 7.

We were jubilant during the last week of school when Mrs. Carlson announced that we would have a clean-up day on Friday. Kids left bicycles at home as they came from almost every direction carrying tools—Fairman Lane to the north, Valley View Road to the south, and down Highway 48. Sam and I brought a rake; others arrived armed with more rakes, shovels, and even a hatchet.

We went to work cleaning and fixing our school grounds, including filling the potholes at home plate and on the pitcher's mound. Behind home plate, boys cobbled together chicken wire to repair the homemade backstop. By afternoon, dirt and gravel filled the holes under the teeter-totter, the craters under the swings, and the path worn around the woodshed. Two older girls wrote a note requesting that the school board fix the broken chain on the swing that had sent them plunging into a mud hole.

Now that winter was over, the water pump worked again. Verna and I pumped a bucket-full, grabbed the schoolhouse broom, and cheerfully offered to swab the outhouse floors with an old rag mop. We swept spiders and cobwebs hanging onto protruding nails from the rough boards of the three-holer toilet.

That's when we heard the screeches. A first-grader had caught his pointer finger in the apparatus of the pump's workings while poking it to examine how it worked, resulting in a painfully crushed finger. We'd learned that lesson once the hard way, too. The little kids who stood by watched in awed silence. Mrs. Carlson cautioned them to keep fingers and hands clear: "And, children, always remember that the pump handle, when brought to its down position, *always* returns to the top, even when your jaw is in its path."

But that incident was the only blight on an otherwise glorious day. Farm kids had learned early to take the knocks of country life in stride; the injured child resiliently moved on to play.

When there was no more work to be done, Verna, Margie, and I perfected our teeter-totter balancing stunt: Margie in the middle, Verna and I on opposite ends, all of us standing tall in perfect horizontal balance. We'd demonstrate our trick for our moms and have our picture taken at the end-of-the-school-year picnic.

Just when we thought this day couldn't get any better, Mrs. Carlson brought out a bag of marshmallows. She lit a match to the heap of dry leaves and grass that had been raked onto the driveway, and we ran in search of sticks; boys pulled jackknives from their pockets to slice off nearby saplings. We speared puffy marshmallows onto willow sticks and toasted them, or, more accurately, blackened them to a crisp over the blazing fire.

With spirits sailing high as kites on our final day of the school year, Mrs. Carlson directed us in tidying up the schoolroom.

Peggy, Margie, and Verna on the teeter-totter at Valley View

We collected and stacked spelling and reading books and carried heavy geography and science textbooks to be stored on the shelves in the back hall. When we cleaned out our desks, to my horror I found a rotten apple jammed way at the back. But it was soon forgotten with the distribution of report cards. We quickly turned them over to look for the words verifying that we were "Promoted."

Then Mrs. Carlson dismissed us for the last time. We were as happy as cavorting calves released onto green grass after a winter confined in the barn as we sing-song-shouted our way home: "School's out, school's out, teacher let the fools out!"

The day after school ended, always a Saturday in mid-May, the whole community gathered for a picnic. By eleven o'clock, parked cars filled every open space in the schoolyard. The softball game was soon underway, with umpire Ed Halberg calling behind the plate. Delicious odors rolled from the schoolhouse. Moms unloaded their picnic baskets onto long tables filled with delights—the usual country potluck and picnic fare. My favorite?

Cool paper cups filled with the Valley View Schoolhouse Punch, made of assorted packages of Kool-Aid, with lots of sugar, and, except for Halloween, always red. Juicy slices of orange and lemon floated on top.

Little Harry and Mrs. Harry, the childless displaced Polish immigrants, were always invited to the end-of-the-school-year picnic, as were the bachelors in the neighborhood. And, while Scandinavians brought food I considered strange, namely those open-faced sandwiches topped with pimento olives, Mrs. Harry contributed a jar of home-canned tomatoes, which seemed even more out of place. Though the neighbor ladies would try to speak to her and include her in the preparation, she was unable to understand and rarely even tried to speak. She would sit quietly by herself until Little Harry decided it was time for them to go home on their tractor.

During a break in the mid-afternoon ongoing softball game, three cardboard cylinders appeared: chocolate, vanilla, and strawberry ice cream! A line formed quickly while moms dipped scoops into double- and triple-decker cones. Ice cream invariably dripped down our arms. Nobody cared.

Before driving home with our families for the rest of the first day of summer vacation, we waved goodbye through cranked-down car windows and shouted our farewells to our friends: "See ya in town on Saturday night!"

METEOR BIBLE SCHOOL

As the days wore on during one summer, Sam and I grew tired of helping with gardening and haying, riding our bikes, picking raspberries, and amusing ourselves in the woods and around Badger Creek. Time yawned before us. So when we were invited to attend Vacation Bible School at the Meteor schoolhouse, we jumped at the chance and looked forward to getting together with

Meteor school kids we did not know as well as with our Valley View friends.

For years, a community church had held gatherings every Sunday at Meteor School. Now, with the hope of filling pew seats in their new building, parishioners and the pastor offered the weeklong program and invited all of the kids in the neighborhood. Dad wasn't keen on leaving his work to drive us the three miles there and back each day, and Mom couldn't. Larry Hawley, the minister, and his wife, Neil, were willing to pick up any kids who wanted to attend.

Mom thought it was as great an idea as Sam and I did. She packed our lunches and we ran to the driveway at the first sound of the Hawleys' car. Three or four Clark kids and Shirley were already packed in, but Sam squeezed in the back, leaving me to sit on someone's lap—something I really didn't want to do. When the kids squished tighter together, an opening appeared. I crawled over a tangle of summer-brown feet and dropped myself into the narrow space. A collective gasp followed, then snickers, as a creepy wet feeling seeped into the backside of my light-blue pedal pushers. I'd plopped my rear end directly into Neil Hawley's lunchtime treat—a freshly-baked raspberry pie. Not to be deterred from his mission, Pastor Larry drove on. After a quick dousing of my pie-stained pants at the school pump, I wore crimson-hued britches for the rest of that day.

At the schoolhouse, Neil divided twenty-five kids into classes for stories, activities, and lessons. At noon, we gobbled our sack lunches while sitting in schoolyard circles, visited, and played games. Then we had afternoon classes. The highlight of each day came when everyone gathered in the schoolroom for Neil's Bible Story Time. Her amazing teaching tool leaned against an easel at the front of the room; it was something we'd never seen before. She pressed her cutout figures (Daniel, the lion's den, and lions) onto the bright green cloth-covered board. They magically attached and stayed put. Neil added a progression of characters and props as she recited the story of the day: sheep, shepherds, even Noah, his ark, and what seemed like *all* the animals. We sat mesmerized, with our vivid imaginations engaged as we watched the enchanting flannel board—the best high-tech teaching tool the 1950s had to offer.

Excitement mounted for our concluding Friday night Bible School Program. All week we prepared by singing "This little light of mine, I'm gonna let it shine," cupping our hands around our mouths, swaying, and swinging our upraised hands in joyous motions. Our hearts were fit to burst. In song, we promised not to hide our light under a bushel. We vowed that we'd (always) keep oil burning in well-lit lamps that outshone the sun. And if ever we built a house, it would stand on solid rock—never sinking sand.

The Bible School Program would be held in the church's brand-new facility halfway up steep Meteor Hill. It would be the unveiling of the Meteor Church basement—and only the basement. In the 1950s, sometimes people would build homes and other structures piecemeal, having only enough money to dig the lower level. They would use it as living quarters, or in this case, worshipping quarters, until they had more funds. Some people lived for years in basements with a flat roof, raised entryway, and a stairwell leading down.

There was only one problem at the newly built church basement: no furniture. We solved that issue right after lunch on Friday afternoon. Like saints marching in, each of us bore our load from school—one chair. In the beating July sun, we paraded in bunches down the gravel road and then turned left to begin ascending the giant Meteor Hill on Highway 48.

That summer evening, from the church basement, children's voices sang out not only on the solid rock of Meteor Hill but anchored *in* Meteor Hill: "On Christ, the solid rock, I stand, all other ground is sinking sand!"

GRADUATION

One year in the late 1950s, five eighth-grade students graduated from Valley View School; Sam was one of them. Folks of every sort filled desk seats. Everyone in the neighborhood, adults and kids, friends and relatives, came with good wishes to send the teenagers off to Bruce High School. Students had filled buckets with trilliums and marsh marigolds to create a festive and fragrant air. But sadness prevailed. That May marked the end of an era in northern Wisconsin: the two white-frame schoolhouses in Town of Meteor would be closed. Like so many other rural schools, Meteor and Valley View succumbed to statewide consolidation.

Although my understanding was incomplete, I already felt a profound sense of loss. My concern heightened when I learned that although Meteor School would become the Meteor Town Hall, my beloved Valley View would be sold—and torn down.

On that balmy May evening, we did not know that there would soon be hurtful disagreements that strained relationships in the

The last graduation day at Valley View. Jane Hanson is receiving her diploma from teacher Mrs. Burhite; seated in front of the chalkboard is Olga Carlson; Mom sits to the right of her, facing the camera.

neighborhood. Most families wanted to join the Bruce School District. But a few wanted to be in Birchwood and made the effort to force everyone there. In our family, Dad chose Bruce and Jerry chose Birchwood. Eventually, all families were allowed to choose their new district, but remnants of the school consolidation clash seemed to blemish community relationships for a long time. It would take serious illnesses to bring our family together again.

Spring wildflowers had barely faded before I learned that Jerry had bought the school building. He took it apart board by board and sold the lumber to John Veness, who planned to build a house in Exeland. Then startling news circled the neighborhood. Mid Jacobson and her husband had heard a dull thud in the night, coming from across their pasture. In the morning, they discovered thieves had climbed to the belfry in darkness and stolen the beloved school bell. That bell had summoned generations of kids to school and hurried us on our way as we looked for trout in Badger Creek or caught bankside frogs. Now the bell was gone.

More dreadful details emerged. Even the outdoor toilets were sold and moved to another farm. There would be no more trips out back to the Boys and Girls outhouses, with wooden seats and holes sized specially for kids. Then the school woodshed, housing extra desks in its attic, was carried away, too. The shed had been where we had shared tales, pranks, and games; where we had played ante-I-over as we chased from one side of the building to the other. One sad summer day as we drove to town, I noticed a building perched crookedly on the hillside in Haw's cow pasture. It looked just like the school woodshed, but forlorn and lonely and all wrong. No kids were running and playing around it. Instead, a herd of cows sniffed at it curiously, as if an alien spaceship had dropped in for a visit.

Slowly the reality of what was happening took shape: my world was being dismantled, piece by piece. Schoolhouse, outhouses, woodshed—gone. And what had happened to the familiar furnishings I knew so well? The gray-and-blue water cooler with the spigot and glued-together lid? The Christmas program streamers and Valentine's Day decorations? The benches, worn from so many years of use, the wrought iron desks fastened together on boards in straight rows? What about the monstrous woodstove with the galvanized metal jacket where we hung our snow pants to dry after recess snowball fights? My favorite *books* from the library bookshelf! Where had they gone? And our set of pull-down maps? The blackboards, erasers, and chalk? Would the wall of our new school have pictures of George Washington, Abe Lincoln, and the Pilgrims celebrating that first Thanksgiving, like we'd had? And just *where were* George, Abe, and the Pilgrims? Where would they be eating turkey this year? When Thorson's pet turkey came to visit school at recess, no one would be there to play! And who would be there for the firsthand education when one of Mid Jacobson's cows gave birth to her calf in the pasture across the schoolyard fence? No one.

Never again would young and old gather together for special events at Valley View: ballgames, Halloween parties, PTA meetings, and the highlights of the year: the Christmas program and end-of-the-school-year picnic. No more delicious odors of chocolate cake, bologna sandwiches, Jell-O, and hot coffee. *No more sugary Valley View Schoolhouse punch!*

That fall, a belching yellow school bus lumbered over the roads we used to walk, run, bike, and sled. We would travel in the warm confines of a vehicle sealing us from the raw pleasure (and pain) of rain showers, snowdrifts, ice, and interaction with the outside world. We would not hear the singing birds, feel the squishy mud on our boots, smell the fresh scent of earth in spring, or stop to taste the oniony leeks. We would no longer watch the flocks of geese in fall, endure the blast of the north wind, or ask our neighbors to warm us when we could endure no more. We would not stop at Badger Creek, or look for sunning snakes, or pick an apple from a roadside tree—or make our own observations.

Armed with new yellow pencils and pencil boxes, boxes of new crayons, tablets of lined manila writing paper, and the improved opportunity of consolidated school education, I entered a world with modern desks, bubblers, indoor flush toilets, fragrant books, and a teacher just for my grade (I was in fourth grade at that point). In the excitement of the moment, it was easier as a child to put away the passing of Valley View School.

The adults were a different story. The welfare of individuals in the community and the local economy suffered. Before statewide consolidation, the day-to-day operation of schools had been a local affair, overseen by each county—in our case, the Sawyer County school superintendent. He dropped by rural schools infrequently to watch the teacher in action and assess the condition of the school. But our two-member school board—Bill Fairman and Luke Birdsill—hired the teachers, oversaw supplies, and arranged for their purchase, mostly from local stores: everything from pencils to sweeping compound, mops, brooms, and kerosene for starting the fire. The board also hired local residents who could use the extra cash. Men cut and stacked wood to heat the school, painted, fixed the roof, repaired the swings, and performed other small jobs. A farmer mowed the school grounds at the start of the schoolyear. Women scrubbed the school from top to bottom over summer vacation and cooked and delivered our hot lunch on every school day. Those jobs for our farming community ended with school consolidation, as did our sense of community.

Yes, everyone still gathered in town to visit and get groceries on Saturday night, and some neighborhood events continued— the showers, shivarees, and holiday picnics. But it was not the same without the Valley View School events to pull everyone together. The heart and soul of our community had also been dismantled. Years later, Dad remembered the sense of loss he felt at Valley View School's closing. But not all of us had fully realized we left a part of ourselves behind.

Going Places

RIDING THE ROADS THROUGH WINTER

Harsh conditions hampered our rural winter travel in the 1950s: bitter cold, wind, and hard-packed drifts of snow and ice. Technology did, too. Unreliable cars refused to start in the cold, had poor or nonexistent heaters, rode on bald tires without tread, and required chains to climb a steep hill.

Extreme winter weather seldom came before deer hunting season each fall, but by late October or early November, the township had hired a crew of local men who could use extra pay—to put up a snow fence. Routinely of faded red, probably painted with economical barn paint, the thin wooden strips of lathe were woven together with wire and flexible enough to be rolled up and stored off-season. After farmers cleared their crops, crews rolled the snow fence into place far enough from the roadside to break the winds sweeping across the fields. Snow collected in massive drifts beside the fences instead of piling up on the roadway. While it worked well most of the time, driving winds or a huge snowfall, or both, resulted in clogged roadways, especially at the top of our hill.

Driving directly into a large drift in an attempt to get through was much like driving into a huge feather pillow—never a good idea. Nevertheless, some people chose to attack a snowdrift this way, sending up clouds of snow. One could only hope the car would emerge on the other side. A drift-busting car was likely to be pulled deeper into the drift, where it remained buried until help arrived to pull it out. Or, just as likely, the car would plunge into the ditch. Driving in the dead of winter was not a sane activity for people without winter driving experience. In earlier times, some folks had surrendered winter driving entirely by putting their cars up on blocks for the season.

At our place, before Dad could drive the car, he had to start it. In extremely cold weather, Dad brought the car battery into the kitchen to keep it warm overnight. He even built a fire in our garage in the stove he called an "airtight." Dad often told the story of someone in a hurry to warm his car who foolishly started a fire under it. The exploded car burned the shed to the ground, too.

We could usually count on getting out in the family car each weekend, visiting with Grandpa and Grandma Walhovd in Birchwood, shopping in Exeland on Saturday night, or going to the Methodist church on Sunday evening. Yet, in our hilly country, Dad always factored in wintertime weather and road conditions. Should we risk driving?

The first challenge was turning the corner over our branch of Badger Creek at the end of our dead-end road to climb the long hill out of the valley. More than once we gave up trying to make it up the hill and stayed home. If we successfully navigated that, going somewhere was easier, since we mostly went downhill, though our return would be challenged by the formidable hills known by the names of the people who lived nearby: the Prilaman hill on Valley View Road, and the Carlson hill at Ole's place. Dad sometimes entirely avoided the icy, never-ending Hanson hill.

On a typical evening return trip home via Valley View Road, our travel scene would have been something like this:

We sailed right along until we'd crossed the dip at Badger Creek by Grandpa Sam and Grandma Bessie's house. Prilaman hill loomed steeply ahead. The car thundered, gathering speed to attempt the climb. We held our breath as snow tires threshed and sent bits of gravel flying, our tires chomping into the snow in the near-zero temperatures. I pictured our car successfully inching its way to the top, creeping slowly to the flat, as I willed the car over the crest of the hill.

We heaved a sigh of relief when we made it, taking a moment while on the heights to look over the distant town below and the wide Chippewa Valley. My eyes scanned across the valley where pinpoints of light marked a farmhouse here and there. In the moonlight, trails of smoke rose up straight from farmhouse chimneys.

Neighbors' dogs yipped as we skimmed on, past the Halbergs, Whytes on the corner, and then coasted down over a small branch of Badger Creek. The headlight beams caught the snowy white coat of a weasel, its eyes glittering as it skittered over the snow bank and into nearby woods.

The car was finally beginning to warm now that we were almost home—Dad always put a large sheet of cardboard in front of the radiator to deflect heat back inside the car.

Our earlier relief was fleeting, however, for before us rose the last and longest obstacle: Carlson hill. This time Dad floored it; the car soared easily over the first rise. But it was too early to claim victory, knowing we must then attempt the final ascent, a short but steep slope just below Ole's house. We crept higher . . . the car barely moving at all . . . nearly to the top. Then the car tires began spinning and we ground to a halt.

Dad muttered, "We'll give 'er another try then."

He shoved the big ball on the floor stick-shift into reverse, backing carefully and avoiding both snow banks and ditches, all the way down to the base of the hill. I tightened my grip on the back of Mom's seat as Dad geared down for another run. This time, the tires gouged through the slick ice and sent chunks of gravel flying. But the road had become icier by our previous passing over it. With tires whirring and whining on the glassy surface, we'd almost won our way to the top. Then the car lurched, slid sideways, and lost momentum. I grew more tense, envisioning our car buried in the bank of snow and the four of us trudging our way home. Maybe Dad would have to climb out and put chains on the tires? I wondered but wisely did not voice that thought, knowing Dad never did.

Mom encouraged Dad with, "I think we'll make it up this time."

"If this don't do it," Dad replied, "we'll have to walk on home."

With resolve, Dad backed yet again to the bottom of the hill and throttled the car to a resounding roar. We held our collective breath once more for what we hoped was the final assault, exhaling only as the car heaved its way to the top and edged its way over the hill. Our headlights gleamed toward Ole's house, his windows sparkling. Then we saw our mailbox at the turnoff to our own home. Unless our short road had drifted closed since we left, it would be a snap to glide on home and safely claim victory.

If it had drifted closed, at least it would only be a half-mile walk.

WEIRGOR

Although winter seemed to last forever, it didn't. But driving during spring snowmelt and summer's torrential rains could also be problematic.

There were actually two routes to Exeland from our farm in the hills, each a trip of five miles. One via Crooked Bridge, a place I'd never thought peculiar until our Indiana relatives pointed out the oddity. They found Crooked Bridge amusing, saying it was the only place they knew where four gravel roads converged and met in the center, with an odd-angled bridge placed smack in the middle. Frequent summer rains sent water plunging down hillsides and turned Swan Creek at Crooked Bridge into a fast-flowing stream. That's when our Crooked Bridge route flooded and became impassable.

At those times, Dad chose the gravel Valley View Road to blacktopped Highway 48, which passed through once-thriving Weirgor. But before reaching Weirgor and the railroad tracks, we had to drive along the edge of the treacherous Weirgor Swamp and through the highway's precarious Weirgor dip.

When Highway 48 flooded at the dip, road crews often placed kerosene-fueled smudge pots to warn motorists. Flaming torches would burn in the squat blackened pots, and the flickering light threw shadows into an overcast rainy day. Oily wisps of smoke smoldered as our car crept along. I watched moving sheets of water rolling away from our tires.

The dip in the road carried further mystique. Although road crews persisted in dumping gravel to fill it, the swamp swallowed up each load and the dip would top with water again after the next rainfall. Rocks, boulders, gravel, and God knows what other kind of fill, all would disappear. Was there no bottom to the endlessly sinking road?

The age-old problem was just as perplexing as the area's name. "'Weirgor,'" strangers often asked, "what does it *mean*?"

Weir, "the diversion of water." *Gor*, of Welsh origin, "excrement" or "deep muck of a swamp." The Ojibwe called it Ba-ke-abash-kang: "a vast and bottomless swamp." The government agreed: on old *Wisconsin Geographic* maps, the area is identified as "The Great Vast Weirgor Swamp."

So, it was no surprise that when companies built logging railroads, they took great care to avoid crossing Weirgor Swamp. At one time, booming Weirgor had been the site of four sawmills in the heart of hardwood country. When Grandpa Sam and Grandma Bessie settled in 1916, only John Halberg's sawmill remained. Still, my grandparents warily skirted the edge of the swamp when they walked the ten-mile round trip to Exeland to get their mail.

The mysterious Weirgor Swamp spawned stories of sleighs lost in the frigid waters, still loaded with perfectly preserved logs. Spinners of yarns circulated outrageous stories of entire carriages and teams of skeletal horses at the bottom of the deep sinkhole. No one quite believed the claims, but who was there to dispute them? Even when I was a child, an enterprising rumormonger might ask, "Heard what th' highway department found 'n th' swamp?"

In later years, when the State Highway Commission worked to improve Highway 48, crews called for massive supplies of rocks and boulders to fill the Weirgor dip once and for all. Locals rushed to answer. They harvested from countless rock piles surrounding every cleared farm field, thanks to bountiful glacial deposits. Farm kids and old-timers alike would never forget the endless work of "grubbing" rocks every spring from fields of heavy clay. During the summer of road construction, farm families depleted their seemingly limitless mountains of rocks and boulders, but all were greedily swallowed up by Weirgor Swamp.

Some claimed that hole went as far as ninety feet deep. No one seemed to know the truth. The dumping of rocks into the dip "once and for all" did not permanently solve the problem: a dip still remains. As does the tale of the Sawyer County dump truck driver who deposited his load of gravel onto the road before passing over the dip. He glanced behind just in time to see the road flip completely over. No one with an imagination and knowledge of local history seemed too surprised. Native Americans, explorers, trappers, traders, hunters, railroad builders, and my grandparents all took pains to avoid traversing that swamp.

Weirgor—Ba-ke-abash-kang. The vast bottomless swamp. An apt name.

GOING TO TOWN

Week in and week out, Sam and I waited eagerly for the trip to Exeland on Saturday night. We knew almost *everyone* would be there. Early each Saturday evening (before Dad bought the used claw-foot bathtub), Mom took down the square galvanized tub from the wall on the outside of our house, placed it in the middle of the kitchen floor beside the cook stove, and filled it with hot water for our weekly bath. I was first, then Sam, then Mom. We emerged squeaky clean and leaped into clean clothes. When Dad finished chores and milking the cows, Mom warmed the thrice-used lukewarm water by adding a steaming teakettle of hot water. While the rest of us busied ourselves in another room, Dad scrubbed to rollicking polka tunes blaring from radio shows on our giant Zenith. *National Barn Dance, Grand Ole Opry*, and the grating voice of Minnie Pearl came through on screechy and static-filled airwaves sent from WLS Radio in Chicago. Once Dad completed his bath ritual, he edged his wicked-looking straight razor across a thick leather strop, lathered the soap in his shaving mug, brushed it on his face and chin, and scraped his whiskers clean. When we smelled the pungent scent of the Bay Rum lotion he'd slapped liberal amounts of across his chin, we knew he was finally ready to go to town.

Whenever we drove, we kept track of other vehicles on the road. Who was going in what direction? We knew everyone's car and truck in the neighborhood and speculated upon destination.

The only car that continually befuddled the neighborhood was Mr. Mikkelson's. He routinely drove oddly designed Studebakers, known for their fronts and backs looking remarkably similar. At a distance, people could never tell whether the Mikkelsons were coming or going, and it became a standard joke. But if it were on Saturday, we knew the Mikkelsons, like almost everyone else in the neighborhood, would be going to Exeland.

EXELAND

In Exeland during the 1930s, a couple of heavy black cars on their way through town occasionally stopped at a tavern on Main Street. Mysterious men dressed in fine, dark overcoats got out, entered the bar, drank politely, treated those in the bar with respect, and sometimes bestowed gifts. Then they returned to their cars and headed north.

"Who were they?" Dad recalled someone asking an old-timer.

"Gangsters from Chicago. Who else would be drivin' such expensive cars, wearin' city folk's clothes, and stop without tellin' us who they are?"

Many locals knew the exact location of the gangster retreat on Blueberry Lake. During the FBI's inquiries into local goings-on, G-men mistakenly tried to find Cranberry Lake, of which there are many in the area. No one bothered to correct the misconception. Although Exeland folks knew well of the comings and goings of those strangers headed to points north and their hideaway near Couderay, Mom actually knew the interior of their lodge.

A relative of Mom's by marriage, Charlie, had been hired as the lodge caretaker in 1928. Not long after Mom graduated high school in 1930, she went to work for him and his wife, Lola. Mom's stint at The Hideout, as it came to be known, was uneventful; as a live-in cabin girl—a cleaner—she reported no sightings of note but did remember the occasional sound of cars roaring in or out during the night (adding credence to the bootlegging booze-running stories that persisted during Prohibition). Charlie's caretaking ended abruptly when he was dismissed after allowing the local baseball team to visit, disobeying strict orders: no unauthorized visitors.

Two decades later on the streets of 1950s Exeland, no gangsters were in sight. Neither was a stop sign nor light. In fact, it was much as it always had been since its founding. In 1907, two railroads met at a point a mile from the settlement of Weirgor. The Arpin Lumber Company logging line, already in existence to extract the rich stands of white pine and hemlock, was crossed by the Wisconsin Central Railroad (which later became the Soo Line) as it extended its passenger and freight line from Chicago to Duluth. The crossing formed an X and led to the community's name, Exeland; in 1920, it became a village.

Forty years later, the railroad depot still served as an anchor to Exeland. Trains brought passengers and deliveries: Montgomery Ward iron heating stoves for the living room; Sears Roebuck furniture, auto parts, farm equipment, and animals arrived via train—even screened wooden boxes of buzzing bees and perforated cardboard boxes of peeping baby chicks. The train was a lifeline, the same as it had been when Grandpa Sam and Grandma Bessie had arrived. And it served as transit to the outside world— should anyone choose to go. The depot brought the best and worst of surprises. Our Indiana relatives came for visits; Joe Platteter, the depot and Western Union telegraph agent, received notifications of death—and sent them. It was to the depot that Dad went to send telegrams of sorrow.

Exeland, with a population always hovering around two hundred, was my first look at a town, any town. At least that's what we called it, despite it actually still being a village. To us, it was simply "town," a place that supplied our needs. But I loved going there to observe the town and those who lived there, knowing I was from the town but not *of* it. Like every small village, Exeland had its share of characters: compromised folks, struggling families, and lonely old-timers; savvy businesspeople, storytellers, and people with brilliant minds hidden behind a veil of ordinariness. Like every small village, Exeland had its share of goodness, meanness, and what we viewed as just plain odd—or interesting—depending upon your perspective.

Exeland might have been the only village of its size that had an Indianapolis 500 racecar driver in business on one end and a Golden Gloves boxer in business on the other. An unusual number of Great Lakes boat captains inhabited Exeland as well. Quite a number of local men took jobs shipping ore from the Port of Duluth-Superior, about one hundred miles away, working on boats for sixty days on, thirty days off; midwinter the men returned to town for months when ice jammed the Great Lakes. Some of them worked their way from crewman to captain, commanding thousand-foot freighters on the always-dangerous waters of Lake Superior.

Our trips to town often included a stop to fill the tank at Taylor's Texaco Gas Station, located midway through Exeland. Harry Taylor, a town leader, solved problems and came to the rescue for all manner of issues—doling out common sense advice as well as twenty-dollar bills to teens who needed prom money. He escorted people over rough country in his big truck, hauled bulky freight from the depot to outlying farms, carted off junked cars from farmers' back forties, and, one time, retrieved Little Harry's "sunny bitch bull" from a well.

Harry Taylor's wife, Annie, always filled our tank, checked the oil, and pumped gas or kerosene into red metal cans for home use. Reaching over to clean the windshield, she babbled as she worked. Annie's high-pitched and always-excited voice also spread the latest news as she bustled about Exeland. She wore a housedress with a flowing skirt, partially covered by an apron flapping in the breeze, and dumpy rolled socks, which she paired alternately either with high-water boots or scruffy bedroom slippers. Strands of her hair always escaped from the scarf knotted around her head. Annie dressed funny, but her heart was pure gold—she continually collected for good causes, selling red paper poppies for the American Legion and collecting dimes for the Red Cross. Annie routinely carried an envelope in which to stash collections to buy flowers for a funeral or give to any family in need.

Taylor's was one of only three filling stations in town; the other two bookended Exeland. On the east, the Phillips Gas Station and repair shop looked across to the Buckhorn Bar, partly owned by the Golden Gloves boxer. On the west, Exeland's twelve-time Indy racecar driver, Frank Brisko, ran Brisko's Gas Station and Auto Repair—conveniently placed next to two more bars where one could wet one's whistle while waiting. When Dad took our car for repair, instructing Frank to "put 'er up on the rack," I watched fascinated as the shiny, smooth hydraulic lift raised our '49 Ford.

In addition to his gas station, Frank ran a small factory in a cement block building. He'd invented Miles Master Fuel Pressure Regulators, a device that promised smooth running, quick starting, and better mileage for modern cars. The town's other attempts at industry took place in a building behind Main Street that reinvented itself for various use. Well before the 1950s, local people used it when they grew cucumbers and sold them to a pickle factory, first storing them in large, round wooden vats of brine before shipment. Later, the building housed a creamery producing butter; later still, a factory building items of fiberglass. But when I was walking the streets of Exeland, it was empty.

Sometimes I'd accompany Dad to Gerber's Feed Mill, located on the tracks near the depot. Farmers would lean against battered rusting pickups while visiting. Cans of Red Man or Scandinavian snus (chewing tobacco) pressed permanent patterns into their front overall bibs; flat cans of Prince Albert pipe tobacco were stuffed in their back pockets. Inside the dusty feed mill, we joined other kids and dads watching streams of ground corn sift into brown gunny sacks. The room filled with the fragrance of molasses, which was added to the grain as a sweet treat for cows. Farmers settled up their accounts, hefted the bulging sacks to their shoulders, and dumped them into sagging car trunks or beds of pickups before driving away down Main Street.

No one I knew ever called the road that went through town "Main Street," though. We didn't call it anything but "the street," despite its being Highway 48. Local artist and store owner Earl Wiggington had once painted the view from outside his store, as if standing in the center of the street, looking west. After seeing his painting, with fresh eyes I observed the road we drove to and from town. After crossing the railroad tracks, the highway headed straight west and climbed gradually into the Blue Hills, then into the Meteor Hills we called the heart of home. The road rose hill by hill, punctuated by a few remaining towering white pines still reaching toward the sky. I suppose it took that painting to make me notice the flats of Exeland in contrast. I could stand on the street and see everything but the churches, school, a smattering of businesses, and Swan Creek, hurtling its way toward the Chippewa River.

Ma Carter's Cafe centered Main Street and the town itself, in a building that had been in continuous operation for decades. Leon and Clara Crye had first operated the red stone café. The scent of grilling hamburger escaped at all times of the day. Inside, six red-topped stools lined the counter, along with a few scattered tables. A faded red pop machine held thick bottles: dewy green 7UP, citrusy Sun Drop, Nesbitt's orange, grape Nehi, and Coca-Cola. Until someone put a dime in, metal bars locked the glass necks of the bottles in place. Customers had to maneuver their chosen bottle into a certain position before being allowed to lift it out, icy cold. Immediately, the metal bars clicked to hold the remaining bottles captive. Returned bottles, in a wooden crate on the floor beside the cooler, would rattle if kicked, purposefully or by accident.

While we waited for someone to hand our double-decker ice cream cones across the counter (five cents a scoop), a dazzling salt-and-pepper shaker collection entertained kids and adults alike. Whimsical glass goldfish, porcelain ponies, and ceramic fawns were displayed in wide glass cabinets. Kids stared in wide-eyed wonder while licking furiously as melting ice cream trailed down an arm. All the while, the rainbow-hued jukebox spun tunes, everything from "Happy Trails" to "Blueberry Hill"—ten cents a song, three for a quarter.

Both townspeople and those living on outlying farms shopped at one of four grocery stores on Main Street. Each store was housed in its own rambling wooden building, two on one side, two on the other. On the south side of the street stood Beise's, with the Wiggington Brothers next door. On the opposite side of the street, Gus and Esther Schwartz owned Schwartz's Store and Meat Market, with Ma Carter's Cafe between it and our favorite store, Veness's. At one time, one of the old buildings had been a rooming house and hotel, complete with a lunch counter that served meals for train passengers and crew.

On the next block, local craftsmen had built the Exeland school and auditorium. Its beautifully crafted red stone matched the American Legion and Auxiliary buildings. Alas, stone walls did not save the school from fire. When it burned down on Christmas Day when I was in sixth grade, it sent adults scrambling to temporarily relocate the lower grades in church basements and vacant places, while my grade bussed to Bruce Elementary twenty miles away.

An earlier auto dealership became Gerber Brothers Auto Sales. Folks from far and wide came for the best deal on new and used vehicles, especially Ford pickup trucks and cars. In my fifth-grade class, we watched in wonder as a young serviceman in full uniform cruised past, driving the spanking-new, two-tone green Ford Fairlane he'd just purchased while on leave.

Down the street, patrons ambled in and out of Dairyland State Bank, a tiny building near the center of town. Ray, a farmer west of town, drove his efficient wife, Edith, to and from work there each day. Behind the teller's window, she penned neat columns of numbers into small leather bank books and doled out cash under the decorative metal cage. I loved to watch when Dad cashed our monthly milk check.

Postmistress Mamie Serley ably managed the US Post Office in a white two-story building. She sorted the mail for daily delivery and sold postal-related products. Mom and I stopped in there to buy three-cent postage stamps and purchase a money order to replace a check for payment to catalog companies.

A small brick house on Main Street housed the Central Telephone Office and its live-in operator. Although most people did not own any telephone, some in town did use wooden crank phones. Kids would gather to watch the switchboard operator seated at her console as she plugged one circuit cord into another to make phone connections. I was so enthralled with the novelty that it became my early ambition to become a telephone operator.

At Eddie Fisher's blacksmith shop just off Main Street, coals glowed over a smoldering fire and leather bellows fanned flames to life. Eddie, whose work always made him look black and dingy, worked behind a protective metal mask. He hammered shards of glowing steel which sent sparks showering as he shaped items at the anvil. Men gathered around on stools to watch while waiting for the repair of a broken plowshare or a mower sickle. Dad warned me not to look at the bright lights of Eddie's work.

The triad of traditional churches, Lutheran, Catholic, and Methodist—our church—were located on side streets. A long-established Mennonite church—historically recognized as the first in Wisconsin—was outside town to the east. To the east and west, respectively, and several miles outside of town, two nondenominational churches sprang up: a community church and the Meteor Hills Bible Church. Seventh-Day Adventists met in local homes.

Other than church on Sundays, weddings, and funerals, there weren't many formal events in Exeland. On Decoration Day, we always attended services at Windfall Cemetery, east of town. Years earlier, someone had planted low creeping phlox in the sandy soil. At the end of May, it was usually in full bloom, carpeting the graves in joyful shades of white, pink, and lavender, adding a cheery touch to an otherwise somber day. The scent of blooming lilacs always hung in the air as an assembly of young and agile servicemen and stooped and shuffling veterans prepared to march in formation. Our neighbors Ole Carlson, Art Hanson, and Hank Halberg, who were American Legion members and World War II veterans, marched straight and proud, flags fluttering in the breeze. Two Gold Star mothers, Elsie Zesiger and Ethel Ploger (who'd lost sons in the war) were escorted to places of honor. Art's command barked, "Ready! Aim! Fire!" Ear-piercing gunshots rang out. Then I waited, spellbound, for Alb Vitcenda to lift his trumpet for the haunting call of "Taps." Slowly, the crowd dispersed, wandering silently toward the graves of family and friends on this solemn day, soothed by the greetings and conversations with people who had returned for the holiday and comforted by the residents of Exeland and the outlying farm families.

Entertainment could be found off of Main Street, as well. For a few weeks each summer, free movies showed at the Exeland Elementary School playground, flickering dimly across a white bed sheet hung between two poles on the hillside. We sat on an old blanket or stretched out on top of the car—if Dad allowed—until we fell asleep or lost the mosquito battle. Occasionally, Village Hall showed backwoods hillbillies or cowboy movies on Saturday nights. Mom laughed uproariously at the antics of Ma and Pa Kettle, but I cried over the Westerns when I thought the cowboys made the horses run too fast and hard. Purchasing a ticket through the hall's authentic glassed-in ticket window somehow thrilled me—Exeland seemed so advanced. And purchasing red and green packaged Nut Goodie bars to munch on was even better.

Saturday night dances in the Village Hall were well attended. Local and regional bands played, such as the Six Fat Dutchman or "Whoopee John," both from Minnesota. After a week of work, farmers and their wives had just enough energy to kick up their heels to polkas, schottisches, and waltzes. Kids listened to the rhythmic beat of drummers, the cheerful *oompah-pah*s of the tuba, and amazing trills on magical accordions and trumpets. I was sure the trombone slide would extend too far and fall to the floor. Although I watched closely, it never did. Mom loved to dance, but Dad had to be enticed to the dance floor, which was sprinkled with corn meal for shoes to slide and glide on.

Most Saturday nights, though, Sam and I followed Mom around to see what went into her grocery cart at Veness's—mostly basic supplies—then entertained ourselves. Sometimes I abandoned Veness's to venture across the street alone to Beise's for a dizzying array of penny candy. Winter, summer, spring, and fall, I clutched my dime or fifteen cents and stood deliberating. If thrifty, I had squirreled away an extra bit of allowance to increase my weekly haul, and I chose carefully from among Cracker Jacks; Saf-T-Pops and Slo Pokes; flavored wax red lips and buck teeth (chew the paraffin, don't swallow); pastel "juice" in wax pop bottles; black licorice snaps, pipes, and black cats; black and red licorice vines and twists; Beemans, Black Jack, Teaberry, and Clove gum; bubble gums, Dubble Bubble (with comics) or Bazooka (with baseball cards); paper sheets of pastel dots; and candy bars of every sort: three-cent chocolate Lunch Bars; pink, white, and brown 3 Color Coconut slices; Butterfingers; and my favorite—swirly Snirkles. They were oblong with circular brown caramel and sugary white "vanilla." Mom and Dad frowned on buying candy cigarettes (they weren't that good, anyway).

In good weather, after purchasing my candy, I met up with

Verna and Richard on the rough sidewalk in front of Veness's (my knees still wear the scars from chasing around playing games of tag). Further down, a metal railing at the top of three steps was perfect for hanging from. The steps led up to the screen door on the Dairyland State Bank, allowing the interior door to remain open for ventilation. We did flips on the monkey-bar-like pipe, but danger lurked just below. A patch of poison ivy stubbornly grew at the side of the tiny frame building. Parents warned their children and expected them to be wary. As far as I know, no adult ever tried to remove it and none of us ever fell into it.

By the time I'd spent my allowance, eaten my allotment of candy (and stashed the remainder for the rest of the week), played tag, and done flips, it was time for the final Saturday-Night-In-Town ritual—possibly the best of all. After dark but before nine o'clock, Soo Line Depot Agent Joe drove from his home on the east end of town to the gold and maroon building at the west end. Older boys and daring younger kids followed the good-natured man to his post at one of the state's smallest Soo Line depot and Western Union telegraph offices. We watched him tap in dots and dashes and clicks, spelling out elation or misery on official yellow

Exeland Soo Line Depot

Western Union paper. No matter how many kids crowded around Joe's desk asking questions, he took time to explain his work.

Even if we'd missed Joe's drive through town to the depot, we never missed the nine o'clock town curfew whistle that also signaled a train would soon arrive, and we had just enough energy to dash toward the tracks. From Veness's we ran—past the fire hall (with its lone fire truck awaiting volunteer firefighters), past Dairyland State Bank, past the village hall, past the school, and past the old white post office. If we arrived in time, someone always had a penny to place carefully on the track to be flattened by the train, hoping to retrieve newly smooth copper. When we heard the shrill whistle of the steam train at the Weirgor–Highway 48 crossing a mile away, we put an ear to the track and listened to the rumbling train approaching town. When red and green lights flashed up the track, we backed away. The thundering train slowed with a hiss, snort, and clunk as the brakeman casually hung by one arm from the steps of the caboose. The engine blew steam as it creaked and growled to a halt. I made sure to look at the cowcatcher at the front of the engine—just in case—then I looked up to see the engineer in his blue-and-white striped hat, hoping he would doff his cap as he peered at us below.

With the train's arrival, Joe sprang into action and noisily wheeled the tall baggage cart and freight cart over the paved area next to the track. He swung bulging mailbags deftly over his shoulder and got ready to toss them aboard the train as someone threw another off. In record time, Joe loaded the freight cart with boxes large and small, fragile and bulky (would my mail order cowboy boots be on this train?).

In the darkness, my eyes darted from one scene to another, afraid to miss any of it—cowcatcher to coach to Joe to carts. Then the magical door swung open and an impeccably dressed black man wearing a neatly pressed uniform stepped smartly from the train, a lantern swinging in one hand. His gold buttons glittered on dark blue wool. His conductor hat sat perfectly on his head, with brass trim gleaming in the night. The passenger coach door rolled open, and he placed a stool for passengers to disembark. They seldom did. Through the windows, unknown faces glowed and peered out into the night. What was there to see in our tiny town? Their stop was short. The conductor slowly rolled the door closed, shutting out the coach's bright lights. With a call of "All aboard!" he gallantly tipped his hat to the group of gape-mouthed kids who only dreamed of ever riding on a train. And then he winked.

The train began to move, chugging and clanging as it steamed slowly down the track until only the red lantern of the caboose was still visible before it crossed Swan Creek on the way out of town. Rounding the bend, it headed into the night, into another world entirely. Then, we turned away—slowly walking back for a week on the farm before another Saturday night in Exeland.

VENESS'S

Each of the four grocery stores in business on the Main Street of Exeland in the 1950s had its own following. Loyal customers seldom crossed the street or went next door to take their trade to another store unless buying a special item only available there (such as my desertion for a prodigious selection of penny candy).

Veness's outlasted them all. True, Archie had died by 1950, but his descendants ably ran the store. Archie's son, Harry, and two daughters, Maude Bartlett and Faye Kellogg, and their respective husbands, Harold and Deb, kept a friendly air, sharing news and cheery information. They greeted us by name as the door hinges creaked and the oft-used screen door slammed behind us, announcing our arrival.

Faintly mingling odors met me each time I entered: ripe bananas, spicy sausage, bacon, bologna, denim overalls, linseed oil, and bins of sweet molasses cookies covered in white icing. My eyes lifted to shelves high above at the front left of the store to notice if they'd put anything new on display this week. Nope. Just the usual: shiny red tricycles, Radio Flyer wagons, galvanized washtubs, large aluminum cookers, giant blue graniteware coffee pots for a crowd, white enamel dishpans with red-trimmed edges, gleaming milk pails, and for-under-the-bed enamel pots with lids—aiding those without indoor plumbing. On the lower shelving stood an array of aluminum pots and pans, stock white dishes, and simple cookware. Seeing no changes there, I examined the small summer toy section containing a few of the usual amusements: school tablets with colorful covers of Hopalong Cassidy, Roy Rogers, and Dale Evans; yellow pencils, small boxes of crayons, and coloring books; red sponge rubber balls, yo-yos, rolled-up kites, and a model airplane or two; games of pick-up sticks, jacks, dominoes, checkers, and boards; paper dolls. At Christmastime, the selection grew larger, including tiny circular wind-up trains.

Mom grabbed a cart, I traipsed along, Sam disappeared in search of friends, while Dad strolled up to the men seated on an odd collection of stools gathered near the cash register. In those pre-telephone, pre-television days, everyone looked forward to socializing when the shopping was done. Dad settled in on a stool and joined the men in good-natured kidding and discussion of the progress of hay or corn crops, weather, fishing, hunting, and local happenings. The men's area of the store reeked of shoe leather, rubber boots, shoe oil, and grease, combined with metallic hardware. Just behind the men, an enormous bin rotated, each metal section filled with nuts, bolts, and washers of every size; six, eight, ten, and every other size of penny nail; harness gear, leather straps, and buckles; staples for every farm need. Shovels, spades, axes, post-hole diggers, hoes, hatchets, and saw blades hung on or leaned against the wall.

Mom pushed the creaking cart through the two grocery aisles, gathering basics for the week ahead. I trailed along, hoping she would pick up something I considered tantalizing—like ketchup. It wasn't likely. There wasn't much to choose from; frozen and snack foods were scarce to nonexistent in the 1950s. Popping your own corn would have to do. I would have considered a bottle of ketchup a triumph, but Mom and Dad didn't consider it necessary in our house. They thought it frivolous to buy anything we could make at home.

Mom filled her cart with ten-pound sacks of Pure Crystal Sugar, Mason jars and Kerr lids for canning, and Sure-Jell pectin—her favorite thickening agent for making jam. In season, we hauled home wooden crates of pears, plums, and peaches. She bought pickling salt; tiny tins of Durkee spices; one-pound boxes of pure lard; and tall, cylindrical cardboard boxes of Quaker Oats for breakfast. Her weekly breadmaking required a cake of yeast and her flour of choice, Robin Hood, which came in fifty-pound cloth sacks printed with floral designs. (She reused the flour sacks as pillowcases. When they were empty, we also reused the one-pound cans of Folgers or Hills Bros. coffee that came with a pry-off key to strip cans open. I added wire handles to make berry-picking buckets. The five-pound cans of Karo Syrup and Aunt Dinah Molasses already came with wire handles, providing more berry buckets.) Finishing off the staples, Mom tossed in a box of Orange Pekoe "Brisk" Lipton Tea for her afternoon tea breaks.

For cleaning, Mom selected boxes of Dreft washing powder for laundry (with a striped dish towel inside the box), powdered Vel dish soap, and brown glass gallons of Clorox bleach. Every now and then, Mom bought a stiff broom or a rag mop head. She made purchases in summer to rid the kitchen of flies: screen fly swatters, small cylinder packages of sticky flypaper, hand-pump sprayers—and deadly cans of DDT. In those days, people routinely sprayed the air inside both the house and the barn with the

miracle pesticide. Farmers and farmwives were thrilled to elim-
inate flies (a decade or so before Rachel Carson's *Silent Spring*,
how would they have known DDT affected all living things?).
Every farm was plagued by flies that clung to the outside of the
kitchen screen door, waiting to sneak through before the door
slammed shut—prompting Mom's summer mantra: "Quick!
Close that door, the flies are gettin' in!"

Mom took the basics to the counter, then browsed through
her favorite area, looking at bolts of flannel and cotton and calico
in bright hues. Bins of rickrack, bias tape, embroidery floss, and
colorful wooden spools of Coats & Clark thread lined the shelves
in a wooden cabinet. Mom would need snap fasteners, lengths of
elastic, and sewing machine needles.

The few unmentionables Mom couldn't sew were placed in-
conspicuously in the ladies' corner of the store: roomy nylon
bloomers, slinky underwear, silky nighties, thigh-hugging "snug-
gies" that reached to the knees for warmth in winter, and rayon
stockings, seamed nylons, and elastic girdles in flat cardboard
boxes. For girls, the awful garters, horrible garter belts, and dreadful brown-ribbed stockings. Storekeepers kept elastic sanitary
belts and packages of Kotex, to attach to the belts, discreetly pre-
wrapped in brown paper and concealed behind the counter. Upon
request, Maude or Faye tactfully retrieved the offending items.

Veness's also had a small section of medicinal home remedies:
cardboard boxes of Epsom salts, blue glass jars of Vicks VapoRub,
oil of camphor for headaches, oil of cloves for toothaches, and
tiny brown bottles of iodine for cuts and scrapes. (Iodine burned
like crazy but felt better when you blew on the wound.) Veness's
also stocked rolls of white adhesive tape for bandages and pink
rubber hot water bottles to chase away a winter chill. The cows
were not long forgotten, so shelves included products used in the
barn as well as the home. The old standby, the square pink and
green tins of Udder Balm, worked equally well for myriad human

ailments. Dads were not forgotten either; this section offered mugs, brushes, and boxes of Williams Mug Shaving Soap (one year, I spent my allowance to buy a bar of shaving soap for Dad's Christmas present).

Year round, every member of the family would be fitted with footwear, from a new pair of Red Wing or Chippewa hunting boots for Dad to boots for the barnyard for Mom to the black buckle overshoes (that were so hard to get off) and black high-topped Red Ball canvas shoes for us kids. In the summer, Veness's displayed stacks of denim wear, both bib overalls (we called bibs) and jeans (we called overalls), summer weight cotton work shirts, Red Heel "monkey" work socks, red patterned bandanas, thin-ribbed sleeveless undershirts (the kind Dad wore), and straw hats for all ages. There were reedy, yellow straw hats for dads and blue or red cowboy hats with strings for kids. (Mom bought me a red one to go with my cowboy boots, but she continued to wear her favorite, an old frayed straw hat she wouldn't replace.)

When Mom and Dad had gathered the rest of the goods, they plunked them on the counter. Faye rang them up on the huge and ancient cash register; a running strip of paper printed out the total. Dad paid by check, cash, or asked the total to be recorded in the "charge book" until he could pay when he received the monthly milk check. Then he packed the filled cardboard boxes into the trunk of the car.

At the end of a summer's evening, after we returned from watching happenings at the depot, Sam and I would often split a Popsicle for a last-minute treat. Deb locked the cash register, sprinkled red sweeping compound on the wooden floor, and with a wide broom swept up the dirt and dust into dustpans, visiting with lingering customers as he did so. While watching him, more than once I still clutched the few remaining pennies from my weekly allowance—to save in my piggy bank—but the coins slipped from my hand and disappeared forever through a gigantic metal grate going into the deep bowels of the mysterious basement below.

One summer night, Veness's wide window caught my eye with a new display: a bright red wagon overflowing with groceries. It included items bearing the Clover Farm label—a pretty pink and green clover, just like what grew in our cow pasture. A big cardboard box, with a slit on top, and a sign announced the contest that would go through the end of July. I excitedly read the directions. A contestant (like me!) should remove the Clover Farm label from purchased items, write her name on the back, and stuff the labels into the big box. At the end of the month, one lucky family would win the wagon and all of its contents! I *knew* I would win. With high hopes, I begged Mom to buy more Clover Farm products, then diligently soaked, removed, and dried Clover Farm labels for weeks, painstakingly printing my name on each to deposit in the box on our next trip to town. Alas, some other happy family pulled away the wagon (such was the start of my run of luck in future contests). Still, it provided a real-life lesson on the odds of winning and a welcome summer diversion while I waited for the next week's Saturday night exploration of Veness's.

LADYSMITH

"Think we'd better make a trip to Ladysmith tomorrow. I'll cut hay in the mornin' and let 'er dry 'til the next day," Dad announced over morning coffee one day in 1957. I mentally counted the money in my piggy bank. Candy or comic book?

Ladysmith, population thirty-five hundred, twenty-seven miles away, was usually a happy outing—unless it included going to our dentist. Because Ladysmith was our nearest big town, Mom and Dad tried to take care of errands or appointments in a trip every few weeks; this was best on a rainy day that halted farm work. Before leaving for Ladysmith, we usually ate an early dinner at home—before noon—so we would not have to eat at a restaurant. Eating out anytime was very rare. If we did, we ate a typical meal at a Ladysmith cafe: plate lunch or hot beef sandwich—white bread, mashed potatoes, and sliced beef, all drenched in overflowing brown gravy. In other words, not much different from the meat and potato meals at home. No one had ever heard of offering a kid's meal. Parents shared food from their plates with the kids.

We sometimes saw our doctor, Howard Pagel, for penicillin for tonsillitis, strep throat, and scarlet fever; sulfa drugs for rheumatic fever; and vaccinations and emergency tetanus shots—namely, for the time I stepped on a nail in a board and punctured the bottom of my foot. Although we didn't like the dentist, we did like Dr. Pagel. He was a very short man who seemed an unlikely soldier, but he had served in the army during World War II and had been highly decorated.

After the doctor, we stopped next at Spidell's Rexall Drug store on the corner to fill prescriptions. Medicinal odors wafted out the front door, and pills came in little blue and white cardboard boxes with flip lids. Then, it was time for some fun, if only of the daydream sort.

On Miner Avenue, through the plate glass windows of J. C. Penney's, I saw displays of all the latest styles. In the fall, perfectly dressed families wore polished saddle shoes and pleated plaid skirts. Well-dressed kids carried matching red plaid school bags. These would not be mine, but, other than the shoes, I didn't want them anyway.

Inside the store, Mom scrutinized tall bolts of bright cotton cloth lined up in rows, removed each to examine the pattern, and then chose appropriately for her next sewing project: school blouses for me. Sam looked for shoes big enough to fit his rapidly growing feet. Dad gathered denim overalls, blue summer work shirts on sale, and white work socks with lines in them—I thought they looked like birch bark. They soon left their finds with Mom and disappeared, bound for different places.

Sam took off for the hardware store to look at jackknives, while Dad went to his favorite stop at James Sporting Goods to examine rods, reels, and fishing baits. Dad *always* chose new baits and fishing tackle. And Sam and Dad never left town without a stop at the army surplus store, still well-stocked in the 1950s. Dad bought lots of well-made, durable World War II goods, all sold at reasonable prices: green rubber "swamper" boots good for the barnyard, khaki-colored canvas pants for duck hunting, heavy-duty knapsacks, and rubberized outfits for the worst rainstorm. Once, Dad even replaced his usual summer straw hat, choosing to wear instead a lightweight safari hat of army green. I thought he looked silly.

At the J. C. Penney counter with Mom, I waited expectantly for as long as it took, for the moment Mom reached into her pocket-book to pay. After the clerk wrapped all purchases, even our shoe boxes, in the ubiquitous brown wrapping paper, she reached up, tucked the money into a small round container, attached it to the wire line overhead, then pulled a cord hanging above her head. The pneumatic tube system whisked the receptacle on the wire to the accountant in the open balcony on the upper level. That woman sat at a desk in the overhanging loft. After removing the money, she stuffed the change and receipt into the receptacle and returned it in the same astounding manner. It zipped into the docking space with a click. This was J. C. Penney's cash-secure system, 1950s-style. I often couldn't stand the wait for Mom's purchase, and, after looking around the store, stood back spellbound near the counter as other patrons paid, never tiring of watching high-tech entertainment.

We did not frequent Ditmanson Co. (a locally owned dry goods department store) very often, but when we did, I would rush to the mysterious fluoroscope—a magical shoe-fitting machine. Standing while placing my foot in an opening miraculously revealed my bones in X-ray skeletal images. My wiggling bones were also magical (the machines soon fell from use, due to concerns of excess radiation).

Sometime during the afternoon, while Dad and Sam were off on their own, Mom and I stopped at the restroom at the Standard Station. It was perfumed with disinfectant. We asked for the key on a chain, Mom spread toilet paper carefully on the seat, and we washed our hands with gritty powdered soap and cold water. We topped off the visit with a drink of water in cone-shaped paper cups pulled from a dispenser on the wall.

Saving the best for last, Mom and I went to one of two five and dimes. At the back corner of one store stood a tall rack with patterned oilcloth for kitchen tables, shelves of draperies, metal curtain rods, and paper lampshades. I watched as Mom chose oilcloth on summer sale and blue plastic bedroom curtains, and then I wandered alone through the incredible aisles.

Down one, I looked at designs on white folded hankies, rabbit's feet on key chains for good luck, and small plastic coin purses. In the toy section, there were plastic farm and zoo animals and tiny dolls with movable arms. I clutched my two dimes and a nickel, perusing Dell comics, a dime each: *Archie, Little Lulu, Nancy and Sluggo, Tweety & Sylvester, Goofy, Porky Pig, Donald Duck,* and *Mickey Mouse.* At the front of the store, I drooled over clear glass bins, seemingly overflowing with their assortments of bulk candy: toasted coconut marshmallow squares, tiny licorice scotty dogs, jelly beans, sugar-coated orange citrus sections, red puffy peppermints, white chunks of nougat wrapped in cellophane, and candy corn; even the disgusting orange circus peanuts looked good. Nearby, some tiny turtles swam in clear round bins while others rested on artificial rocks or under unimaginably fake green palm trees. Twittering parakeets, feathered in green, yellow, and blue, flitted happily about in a gigantic bird cage and chattered on swings while admiring themselves in mirrors. Bird-seed and feathers flew in all directions.

Mom caught up to me. Overcome with all the delights one could purchase, I pleadingly asked the question I already knew the answer to: "Mom, just *once*, can't we buy anythin' we want?"

"Someday," she said, "when you're grown up. Someday, when your ship comes in."

I settled for a *Nancy and Sluggo* comic book and coconut marshmallow squares.

Once while Mom shopped, she let me go on my own to what I hoped would be the very best part of my day. I climbed the steps of the stately stone Carnegie Library and entered through a tall, heavy door; dark shelves towered over a wooden checkout desk in the center of the room. I stood gazing in wonder at what seemed like an unbelievable collection of books. Moving toward the back in a dreamlike state, I located the children's section. At least a dozen Nancy Drew books, my current favorite, filled a long shelf. How I longed to carry an armload of books down those tall stairs! Finally, I gathered the courage to ask the librarian behind the massive desk, "I'm from Exeland. How can I check out books?"

"Oh, this is a Rusk County library," the librarian informed me politely, peering down from her looming desk. "I'm afraid you're from Sawyer County. You aren't eligible to check out books here."

Quiet overwhelmed the already quiet room. With that, the librarian turned back to her work and I walked away dejected, allowed only to look. Mom had warned me the library would probably not let me check out books, and she'd been right. For now, until the free library books Mom had ordered for me arrived in the mail, the comic books I'd bought at the dime store would have to do. In future, I would return to note new titles in the Nancy Drew section.

Finished with our shopping, our family of four would meet at the car, often parked outside Dr. Pagel's office, and pile in for the ride home. If we'd been gone most of the day, sometimes Mom hadn't had time to bake fresh loaves of bread for the week or thaw out a package of meat for supper. Then Mom asked Dad to stop in Exeland at Veness's. I hoped she'd pick up already-sliced, white Bunny Bread; I heartily agreed with the store-bought bread jingle: "That's what I said, Bunny Bread." A ring of bologna and a can of Van Camp's Pork and Beans rounded out my favorite meal. Sam's, too. Then he and I fought over who got the only greasy piece of pork in the entire can.

Most trips to Ladysmith ended in satisfaction.

DOC DISTRESS'S DENTAL OFFICE

Mom's medicine cabinet was well stocked with alcohol, iodine, gauze, and adhesive tape—but never Band-Aids. She considered them frivolous. All we needed were Epsom salts for soaks and plenty of medicated rubs, as well as Rawleigh's ointments and soothing salves—the traveling salesman brought those to our door. Mom also made sure we had the essentials of tooth and oral care: lumps of alum for canker sores and oil of cloves for toothache. But a stubborn toothache could be relieved for only so long. Then the dreaded trip to Doc Distress in Ladysmith became a necessity.

In Dad's eyes, Doc was okay. First, he had been around for years and had a name known to Dad. Second, he kept his prices low, having changed little in eons. Third, Dad couldn't resist a visit to conversationally scope out Doc's fishing haunts. I soon learned Doc Distress was better known for his fishing skill than his dentistry.

The ascent up the creaky wooden stairway was long and foreboding, with an overpowering scent of everything old mingling with customary medicinal odors. It assaulted the nose even before we entered the second floor. In his day, Doc Distress may have been a crack dentist, but those days were long gone. Certificates on the wall proclaimed him to be a graduate of the Marquette School of Dentistry, though the aging yellow tint of the papers lent some question to his credentials. They had most likely been issued at the turn of the 1900s.

Doc had no receptionist or assistant, causing further reason for concern. A thick layer of dust across the chairs made it look as if it had been weeks, or perhaps months, since another patient had crossed this threshold. Dog-eared fishing magazines cluttered the lone table, as if Doc Distress had a lot of downtime and plainly did not spend it reading professional dentistry journals.

Everything was ancient, especially Doc. His weak and crackly voice droned on as he shuffled feebly about the office, readying his antiques in preparation. Evidently, he spent his available income on bass boats and fish killers rather than updated dental equipment. Finally, the torturous tools were ready, and because I was a trusting and compliant child, I climbed into the groaning, decrepit chair.

Doc Distress was not unkind, but had no "chairside" manner either, preferring to ignore me while he worked and discuss fishing adventures with Dad. While Doc talked lures and lines, he drilled, his voice accompanied by the endless whirr and whine of aging equipment. I stoically endured the pain and the sickening smell of hot tooth enamel rising from my mouth, only lifting my head obediently to rinse and spit into the tiny enamel bowl when commanded.

Did I mention that Doc thought the use of Novocain unnecessary? He did not believe in X-rays, either.

Apparently, Doc's eyesight was not so good. More than one of my teeth became painfully abscessed because of improper treatment. When I returned to have one tooth extracted, the man could barely manage the task. While Doc Distress had skillfully mastered pulling hooks from the mouths of Lost Lake fish, extracting a molar from my mouth proved more difficult.

On the bright side? With multiple trips to Doc Distress, at least Dad had gleaned prime fishing tips.

GRANDMA AND GRANDPA WALHOVD'S BIRCHWOOD HOME

Before 1956, when the State Highway Commission paved Highway 48, we drove to Grandma and Grandpa Walhovd's on a dirt and (sparse) gravel road. The dust rose up in gritty clouds as cars rolled over it. In the wake of a lone passing car, we'd tightly roll up our window until the worst had cleared. But dust seeped through anyway on the Old Stony Hill Road, named so for good reason. The rough and rocky road often resulted in sudden carsickness for Sam or me. Sometimes Dad had to pull over for one of us to get sick in the ditch. Yet, it was an interesting drive that meandered through heavily timbered woods with tall trees on either side. It was not unusual to spot slow-moving porcupines, tawny-red roadside deer, and partridges drumming on logs. Dad took more or less the same fifteen-mile route he had taken when courting Mom for those seven years.

As we neared Birchwood, I loved thinking about Mom's childhood stories. As we crept down Breakneck Hill, I visualized her sledding parties; as we passed Drinkwater School, I envisioned Mom and her friends taking sleigh rides to dances.

Finally, our car rolled onto Cedar Avenue in the upper part of Birchwood; my grandparents' home on that street was easy to spot because a giant oak tree towered over the front porch. The tree was so big that Sam and I together could not get our arms around the trunk.

Grandpa and Grandma Walhovd's house was a veritable spice

Grandma and Grandpa Walhovd on their wedding day

box of fragrance, starting with the entrance through hanging pots of blooming petunias, their sweet bouquet filling the front porch. Grandma, ever-busy, gave hugs through her wet apron, then hurried to the kitchen; Mom followed her, chatting away. From the kitchen, I smelled a peppery scent mingling with the rich earthy odor of the packed dirt basement floor below the dining room. A whiff of Grandpa's pungent pipe tobacco drifted from his pipe rack and the green glass humidor that rested on the library table alongside his rocking chair. A lingering trace of Grandma's talcum powder wafted from the corner bedroom off the parlor. Years later, when I inherited Grandma's small writing desk, an open drawer still retained a trail of her talc.

Grandpa, a tall, solemn, full-blooded Norwegian with a bald head except for the fringe of white hair above his ears, greeted us quietly, then returned to his chair or his work at a tall dark desk he called "the secretariat." He invited Dad to sit and visit before Dad (usually) departed for fishing lakes north of town.

Sam and I joined Mom and Grandma in the kitchen; a crank

coffee grinder hung on the wall beside a wooden spice cabinet with tiny pull-out drawers. Grandma's cookie jar was loaded with toothsome treats: crisp ginger molasses cookies, buttery date-filled cookie crescents, and my favorite, fattigmann bakkels—Norwegian poor man's cookies—fried in lard, sprinkled with sugar, and crunchy to the bite. Although Grandma was all English, she had learned to make Norwegian cookies and dishes, including a reputable plate of lefse.

After Dad left, we had coffee and cookies. On the center of the dining room table sat a jar for spoons, a vinegar cruet, and a sugar bowl. Much to our delight, Grandpa always piled the sugar bowl high for our visit. When he dropped a lump into his coffee cup, Sam and I popped one, or more, into our mouths, too.

If it were late spring or summer, we knew we would not be the only visitors at my grandparents' house. As a descendent of generations of women who had survived by their strong independence, Grandma had an ingrained entrepreneurial streak. It had been honed by raising eight kids on Grandpa's carpenter's wages through the Depression, a challenge that Grandma had taken on with energetic zeal. She had always looked for the best deal and found ingenious ways to earn a little extra money, something she still did.

Sunday afternoons brought a regular stream of locals dropping by Grandma's house. Birchwood folks knew her as their source for plants and flowers. She was especially busy before Decoration Day and during planting season in June. With her trowel, Grandma carefully lifted plants from wooden flats; their tiny root clumps had plenty of soil still attached: green pepper plants, tomato plants, and starter flowers. She rolled each purchase into its own small cone-shaped newspaper bundle for customers to transplant in their gardens or at a grave in Woodlawn Cemetery.

Grandpa had built a sizeable greenhouse in the sunny front yard where Grandma started plants from seed. As an original

Grandma Walhovd's greenhouse

organic gardener, she heated her greenhouse with a small wood-burning stove and reused the ashes to complement her soil. She also buried fish, fish cleanings, and food scraps, later mixing all to fertile perfection in soil creation. Her huge garden overflowed with flowers and produce. Like Mom, Grandma stored away the fruits of her labors on basement shelves: homemade ketchup, watermelon pickles, piccalilli relish, applesauce, ground cherries, jams, jellies, and preserves.

Customers also came to buy Grandma's worms. Sam and I watched as she pushed a simple wire device containing electrical current into the ground. That brought night crawlers and angle-worms to the surface. In the lake area, summertime tourists were always in need of the bait that Grandma sold for ten or fifteen cents a dozen.

When Sam and I tired of seeing people we didn't know, we ran off to examine each corner of Grandma's sizeable yard. We chased through the rows of flowers, looked over her vegetable plots spanning the long sloping yard, and peeked in the green-house. In the backyard, Grandma had an honest-to-goodness cellar door, slanted and perfect for sliding down—but beware of splinters. Grandma's rain barrels were situated at each corner of the house. When they were empty, we hollered down those barrels and listened to the echo of our voices. When they were full, we splashed each other and dipped our plastic water pistols full, then chased and shot at each other as we dashed through the yard.

Next door to Grandma's greenhouse lived John and Doscia Monteith. Doscia, a tall woman with a peculiar name I'd never heard before (pronounced *dos*-key-ah), wore a faded apron wrapped around her housedress. She seemed always in a hurry and often rushed over to speak to Grandma in her heavy European accent. The two women had bonded through the many decades, over children, housekeeping, gardens, and husbands. In 1951, their marriages were just two of the six in Birchwood reaching the fiftieth anniversary milestone, the same year Birchwood itself celebrated fifty years.

Grandma Tuttle lived on the other side of Grandma's, in a little square house on the corner. One Sunday, Mrs. Tuttle's teenage grandson, several years older than I, introduced me to a game he called mumbley-peg. The game seemed interesting enough, but I didn't own a jackknife. Grandson kindly loaned me one of his, snapping open the long blade as he explained the rules and some system of scoring points. Demonstrating how to play, he stood with his feet shoulder width apart, seized the tip of the extended blade, and hurled the knife end over end. The blade plunged into the ground right beside his canvas-clad foot. He scored points for proximity. Then he handed me the knife and invited me to do the same. Next to my foot? My fumbling attempts didn't even come close; I didn't win, nor did I care. I didn't tell anyone about the game, not even Sam, and never tried it again.

Although Grandpa and Grandma had an indoor bathroom, they still occasionally used the Walhovd family outhouse, especially when working in the garden or workshop. At the side of the lot, a brushy thicket of wild plum trees discreetly screened it, shielding it from Mrs. Tuttle's view. The plum trees produced lovely pink blossoms in springtime and plump juicy plums in fall.

Beyond the outhouse, at the far end of the lot, stood a vacant chicken coop, a garage for Grandpa's black Model A Ford, and Grandpa's carpentry workshop. Grandpa passed the time making rocking horses, doll-sized beds, tiny cupboards for my play dishes, and corner cupboards and decorative shelves for Mom. I often eyed my uncle Vic's long-unused catcher's mitt hanging on the workshop wall. I always wanted to ask if I could have that glove, but I was too shy to ask. Besides, I didn't want to be a catcher, so what would I do with a catcher's mitt?

Uncle Vic lived with Aunt Esther in a small gray house behind the back yard. Grandma would summon them over for coffee by waving a dish towel out the back door. Three of Mom's six brothers had each married a woman named Esther, so the whole extended family had to numerically identify them. Since Vic was Mom's oldest brother, we called his wife "Aunt Esther One." Until she had it removed, she was also the aunt distinguished by a huge goiter growing under her chin. Once, all three Aunt Esthers lined up for a photo, each holding the number of requisite beer bottles corresponding to her title. A six-pack of Breunig's Lager did it.

AROUND BIRCHWOOD

The Chappelle house stood directly across from the Walhovds' and overlooked open country that everyone called the Range—land originally owned by lumber, railroad, and cattle baron Frank Stout. At his death in 1927, Stout was considered one of Chicago's wealthiest men. Knapp, Stout & Co., founded in part by Frank's father, had accumulated massive wealth from harvesting the native pine of northern Wisconsin, leaving the cutover from which grasslands then grew.

In the early 1900s, Stout then shipped by rail thousands of sheep from Montana. Doscia's husband and another man were hired as shepherds. Shearing time kept five experts from Montana busy for three weeks, even with using gasoline powered clippers. Then Stout turned loose hundreds of Hereford cattle to graze on the plentiful grass. Was it any wonder locals called this area outside town "the Range"?

One Sunday afternoon, I clamped metal roller skates to my shoes in an attempt to skate on the novel blacktop surface available at Grandma's house. When the pocked roadway proved too rough, I took off my skates and gave up, just as Herb Chappelle emerged from his backdoor wearing his pin-striped bib overalls. Every Sunday afternoon, like clockwork, Herb donned his overalls and hat, tucked a cigar box under his arm, and walked two blocks over to the baseball field. Some Sundays, Sam, Mom, Grandma, and I dropped by to watch the Birchwood team, too, where we would find Herb dutifully collecting money by passing the cigar box through the bleachers, urging fans to drop in a few coins in support of the home team.

Right across the street from the bleachers, I could see the red-brick two-story house Grandma and Grandpa had lived in when they first moved to Birchwood, when Mom was six. Grandma said they briefly rented the home from James Morey, a prominent village businessman who had held positions as Birchwood's postmaster and director of the Birchwood State Bank. He had also owned the sawmill and lumber yard. According to the *Rice Lake Chronotype*, one of Morey's sons had ridden his pony to round up Frank Stout's steers for shipping to market. One of Morey's grandsons went on to become a giant in the aviation field, establishing airports around Madison, one of which remains Morey Airport, in Middleton. The red-brick house later became the Birchwood Historical Society Museum.

Just behind the ballfield, Mr. and Mrs. Shimanick owned the blacksmith shop—dark, dim, and coated in soot. Mr. Shimanick seemed impossibly old and frail, but all the same he pounded on hot steel and sent sparks flying. Sam and I watched in fascination while Mom and Grandma visited with Mrs. Shimanick, then we continued on to Mom's childhood friend she called "Buzzy." She lived across the street from the school playground, our next stop.

No Sunday walk about Birchwood was complete without the novelty of playing on the merry-go-round and slide. The first time down the slide, I was afraid to release my iron grip at the top of the slippery metal. Sam pried my fingers loose and I sailed to the ground, instantly changing my mind.

Walking the block to Main Street, the four of us would turn left toward the Birch Lake Hotel and Bar, which anchored one end of town and overlooked Little Birch Lake. This was when Mom's stories really came alive. As we walked, I imagined Mom's ice-skating parties on the frozen lake and gypsies rolling into town. Where the Red Cedar River tumbled over the dam, in the area known as Sleepy Hollow, a bait shop stood next to a small sandy beach. Some swam (riskily) near water flowing over the dam that, in Mom's childhood, had produced the town's intermittent electricity.

The river ran past Eddie Weise's Riverside Dairy, the source of the metal crates of milk delivered daily to Valley View, Meteor, and other country schools. The dairy bottled milk from local farmers, sold it locally, and also ran its route to deliver the half-pint bottles to schoolchildren.

Opposite the dam, on the other end of Little Birch Lake, the Birchwood Lumber Mill still had stacks of logs (although Birchwood owed its lively existence to the railroads, it began with the creation of the mill and a thriving wood business). Walking with Mom back up to Main Street, I saw the Omaha Railroad skimming through the edge of town and its depot at the end. In my mind, tramps and hobos gathered around a campfire with their stew and tin cans.

Mom and Grandma had something better than hobo stew in mind for us.

Along Main Street, we stopped for ice cream cones at the restaurant and bar known as the GATE, so-named for the owners at the time: Gib, Al, Thelma, and Esther. The GATE was in the longest building I had ever seen; it was divided right down the middle—restaurant on one side, bar on the other. The restaurant included a soda fountain furnished with old-fashioned, wrought-iron and wood ice-cream-parlor-style tables and accompanying chairs. There were even miniature tables and chairs for kids. For five cents, I purchased tiny cellophane boxes of salty Rold Gold pretzel sticks, counting out equal sticks for us to share.

On some Sundays, Mom, Sam, Grandma, Grandpa, and I climbed into Grandpa's old-fashioned Ford Model A, circa 1929, and put-put-putted the four or five miles around Big Birch Lake. Grandpa seldom spoke, but occasionally he'd point out a house, cabin, or building that he'd built. Over the years, he'd built dozens, even hundreds, all around the Birchwood Lakes area.

Aunt Bert and Uncle Lester, one of Mom's twin brothers, operated Echo Bay Resort on Big Birch. Our six good-natured cousins made full use of the lake, swimming, fishing, or rowing boats out to Penny Island. Each swam like a fish, putting Sam and me, fearful landlubbers, to shame. The jolly cousins seemed to have plenty of candy and comic books to share. We saw them some summer Sundays and also when all of Mom's relatives gathered at Echo for a family reunion. Mom's Janesville relatives would bring a carload of contraband with them, in the form of inexpensive oleo—margarine—technically illegal in Wisconsin at the time. My penny-conscious grandparents ate it, but Dad, being a dairy farmer, refused to eat anything but butter.

On rainy days, Sam and I stayed inside and ventured up Grandma's steep stairs to a low-ceilinged room. It still held the childhood row of narrow, metal-frame cots for Mom's six brothers. Grandpa had added to the original boxy house room by room, as money allowed, to accommodate the family of eight kids. A black-and-white photograph of Old Abe hung on the wall. The image wasn't of the nation's sixteenth president; it was the bald eagle that had served as mascot for the Eighth Wisconsin Volunteer Infantry Regiment in the Civil War. It hung in tribute to Grandma's grandpa, Simon Peter Gansell, the man who had died of disease in training and left his wife to fend for herself and two children. Old Abe reminded me of my great-grandma Phoebe's Civil War orphanage upbringing and her Chautauqua days when she would sometimes bring her young daughter (my Grandma Walhovd) on stage with her.

When musical tones drifted up the stairs, Sam and I would be drawn to the living room, where Grandma sat swaying on the piano bench while playing ragtime. A rack beside the upright piano held stacks of sheet music, although she didn't need them. Like her mother, Grandma could read music, but she played by ear. Over the years, she'd played with Saturday night dance bands and at silent movies in Birchwood. Sam, Mom, and I rocked back and forth listening as we sat in padded, Mission-style wooden rockers.

Before long, Dad would return from fishing. He and Grandpa shared fishing stories and exchanged old hunting tales. This they did, occasionally, over a taste of Grandma's homemade wine. Though it is unlikely a drop of alcohol ever knowingly passed my Grandma Bessie's lips, Grandma Walhovd occasionally uncorked a bottle of her rhubarb, grape, elderberry, or dandelion wine.

Then came our last rituals. The six of us, and sometimes Uncle Vic and Aunt Esther One, gathered for a meal around Grandma's supper table. When she finally sent us on our way, we left carrying cookies or leftover boiled spice raisin cake that Grandma said Grandpa "wouldn't eat," funny papers she'd saved for us, and riches from her greenhouse, garden, or cellar shelves.

Armed with tangible biweekly treasures and my intangible observations of a town unlike Exeland, we drove before darkness fell into the Meteor Hills toward home, hurrying to bring the cows in from the pasture, finish the evening chores, and await our next visit to Grandpa and Grandma Walhovd's house, where Mom's stories came alive.

Freeman's Favorite—Boiled Spice Raisin Cake*

2 cups raisins
2 cups hot water
1½ cups sugar
½ cup lard
Spices of choice (at least 1 teaspoon each of cinnamon, ginger, nutmeg, and allspice)
1 teaspoon salt

Boil the above ingredients 3 minutes, then cool to room temperature. Add 4 cups of flour, a teaspoon of baking soda, and a teaspoon of baking powder. You may also add chopped nuts. Pour into a large 9 by 13 pan. Bake at 350 degrees until done. May be frosted with caramel or maple icing.

*The recipe has various names; likely it was introduced during the lean World War II years. Note that no eggs or milk are needed. And it happens to be my favorite, too.

WINDFALL LAKE AND SPOONER

Our dips in Badger Creek were good enough for cooling off, but on steamy summer days, Sam and I longed for a swim at Windfall Lake, a mile and a half east of Exeland, where people gathered on a narrow sandy beach along County Highway D. No, we couldn't actually swim, but we loved splashing, wading waist high, and seeing our toes wiggle in the sand through a crystal-clear Wisconsin lake.

We didn't know how to swim because no part of Badger Creek was deep enough for that, we were scared of water over our heads, anyway, and neither lessons nor a swimming pool were anywhere to be found. So we remained non-swimmers, wary of the Windfall Lake drop-off a short distance from shore. With no lifeguards or markers of any kind, it was up to kids and their parents to be watchful. Once, when I stepped into watery nothingness, unable to plant my feet on sand, I was terrified. I emerged, glugging water and with a stinging nose, quickly looking around to see who had noticed my fright and quietly vowing to be more careful. Once in a while, Mom, who could swim, donned a suit, too. However, she didn't attempt to teach us how to swim on our infrequent journeys to the lake; I suppose we just wanted to play in the water— or, maybe, Mom simply wanted to stay submerged in her faded red one-piece suit with the stretched-out nylon that had been given to her by a friend?

One day, Mom uncovered a man's one-piece suit from days long gone. Hidden in the attic, the scratchy black wool bathing costume was a 1920s affair, strangely covering the chest, as well. I'd never seen anyone wear such a thing, not even in pictures, and I was horrified when Dad (who couldn't swim and had never even been seen in shorts) decided to wear the holey moth-eaten suit and go in Windfall Lake. I thought Dad must have suffered sunstroke. Dad's never-seen-the-sun skin seemed to glow ghostly white in the evening light. I'd never been so embarrassed. And I couldn't wait to go home. Thankfully, the suit disappeared, and Dad never repeated the episode.

Besides those rare trips to Windfall Lake, we waited all summer for three more glorious days, saving bits of our allowance for each: the rodeo, the lumberjack show, and the county fair. All amazed. Sam and I waited with the intensity only the young can experience . . . certain the magical days would never come.

The Spooner Rodeo would be held in early July, and, yippee!, my feet would still fit into my hard-earned cowboy boots. Mom had sewn me a pale red denim cowboy shirt, trimmed with bright red corduroy cuffs and pockets. Unfortunately, she embellished with a small decorative collar of the same bright red material. I noted that real cowboys didn't wear fancy collars—to no avail. (Mom still hadn't given up trying to make me ladylike.) But I made up for it with the gray pants I thought looked *very* cowboylike, since they had a single pearly decorative snap on each front pocket.

My Western wear was enough for me to imagine myself sitting alongside the cowboys on the corral rail. Me, among handsome cowboys. I'd slide into the saddle of a bronco as he bucked his way into the ring, gate swinging wide, and stay aboard for the entire ride around the ring. Later, I'd be one of the gallant cowboys riding to scoop thrown riders to safety behind my saddle, deftly guiding my horse away from the snorting Brahma bull.

If only I had a pony . . . I would rope steers and hog-tie them into submission before the cheering crowd going wild at my record-breaking performance. Swaggering cowboys would gather 'round in admiration as I was awarded a sparkling silver belt buckle. Then, I'd be whisked onto a shining stallion (by the best-looking cowboy) for a victory lap around the ring, and, waving my cowboy hat at my adoring audience, I'd stand in my stirrups and bow to the cr—

Snapped back from my fantasy world by Brownie's bellowing as I sauntered along with our cows from the pasture, I sighed at the holstered six-shooters hanging limply from my waist; they were fake and my cowboy hat was made of straw. Still. I turned up the brim of my straw hat in determination, just as I'd observed the most handsome cowboys did with theirs.

Finally, the day arrived. After our Are-we-there-yet? thirty-seven-mile drive to Spooner, the rodeo kicked off with a grand parade of riders circling the ring. They bore colorful flags and were accompanied by the local saddle club: adults, kids, and yes, even girls, all on horseback, smiling widely. Long slinky fringes dangled from their sleeves, and felt cowboy hats of red, black, and blue either hung loosely across the rider's back or were held on by ties under the chin. All of the girls rode easy in the saddle, even as the horses broke into a canter, then a smooth gallop. From our bleacher seats, I thought I could hear the squeak of shifting leather saddles, jangling spurs, and the slap of leather chaps.

Emboldened by those girls in the saddle, I tried my luck: "Okay if ya'll set here for a spell, Mom? I'm headin' over yonder ta say howdy to a dude over that-a-way 'fore he saddles up 'n' heads out ta—"

No? Oh.

Just as the bronc saddle riding, bareback riding, calf roping, steer wrestling, barrel racing, and rope twirling show began, the balloon man passed slowly by, selling outrageously priced trinkets. I craned forward, half-standing, to admire the carved leather belts, real leather chaps, and a new set of holsters. Mom quickly settled me back into my bleacher seat and was further saved by events in the ring.

A sagging jalopy bounced along, swirled in a snappy circle in the dirt, then skidded to a stop. One creaky car door swung open, and the driver emerged wearing long floppy yellow shoes, silky ballooning pants, a polka-dotted red and white shirt, a shocking green bowtie, and the most gigantic cowboy hat I'd ever seen. Strutting forward, he straightened his tie and adjusted his hat. But my attention was turned back to the car. The trunk popped, the hood was thrown open, and from every door came more clowns—each more outrageous than the last. Bumbling stupidly to action, they leap-frogged and tumbled over one another, each antic more ridiculous than the last. I collapsed in senseless laughter.

And then, it happened!

Gates swung open suddenly and released an angry bull twisting in every direction as its rider clung to a short rope. When the rider was thrown to the ground, the bull turned eagerly to stomp the cowboy into dust. That's when the clumsy-looking clowns appeared again, teasing and distracting the pawing, snorting, agitated animal. But the agile clowns were too fast and kept just a step ahead, sometimes diving into the safety of barrels, which only further infuriated the bull. While the bull rolled barrels with its head, the cowboy safely escaped.

That unusually hot summer afternoon, we sweltered in the direct sun. Though we were summer-tanned by then, Mom feared we were too hot and summoned the balloon man to buy Sam and me Asian-style conical hats. Mine was shocking pink, Sam's was vibrant green—far cries from authentic cowboy hats.

Happy but worn out, I gathered the cows on foot that evening, sans horse. Still star-struck, for a few days I dreamed cowboy dreams, then soon turned my mind to the day a few weeks later when we'd go to the Lumberjack World Championships in Hayward.

HAYWARD

Once a (very) bawdy and rough logging town, it was appropriate that Hayward celebrated its logging history according to the northern Wisconsin slogan, "Hayward, Hurley, and Hell." (Hurley being another notorious logging town. Hell being, well, Hell.)

Our afternoon began on bleachers overlooking the competition area in and around a narrow bay in Lake Hayward, now called the Lumberjack Bowl. For the log rolling, or "birling," differently sized logs had been tethered afloat in the water, and their colored stripes indicated size and difficulty. Alongside the bowl, huge chopping blocks waited for lumberjacks to compete with axes and saws, according to various styles of both. On the opposite side of the bowl stood sixty- and ninety-foot poles, ready for speed climbing. Male—and even a few female—competitors strolled about, dressed in buffalo check shirts cut at the elbow for summer wear and high-water pants with suspenders, just like I'd seen worn by the old-timer lumberjack who'd visited Valley View schoolhouse. A small paddleboat called the *Namakagon Queen* plied the waters dock to dock to pick up and drop off competitors.

With introductions, the show began with contestants from the United States, Canada, Australia, and New Zealand. One year, a neighbor boy competed in the chopping contest. He was strong and muscular but no match for the professionals who knew the strategies.

Dad and I were especially entertained by the birling competition. Men, women, and even children wore spiked boots to grip floating logs as they spun them, one person at each end of a log. To take control, each bounced and attempted to spin the logs first one way, then the other, trying to cause the opponent to lose balance and tumble into the water. The secret? Never take your eyes off your opponent's feet. Even that didn't always work; Dad chuckled heartily when one of the birlers slid into the frothy drink. I thought of old Hank Hendrickson's visit to our schoolhouse, his hobnail boots and high-water pants. With hobnail boots and a little practice, could I do this too? Maybe Dad could float a log in the pond north of our house? (I didn't even ask—not when I thought about how mucky the pond was.)

Dad was fascinated by the speed chopping, bucksawing, and the men who scrambled up the towering poles to ring a bell at the very top. They were secured only by a rope thrown around the pole, and it took them just seconds to ring that bell. Those events rekindled fond memories for Dad of his logging days in the woods he loved so well. Back in the day, one of his logging crew buddies had gone West to work in the really big timber of Oregon. He returned boasting of climbing ninety-foot trees and using a springboard method for chopping massive pines.

As we left the grounds, heading for our hour-long ride home, Dad had a faraway look in his eyes, though he made no comment. A day or two later, Ole stopped in for a visit, asking, "How was that there lumberjack show, then?"

Dad, who never said *ain't*, reverted to lumber-camp speak, "Well, it ain't the ol' loggin' days, no more, Ole. Now ya git prize money for doin' the hard work we once did ta make ends meet."

Maybe Dad felt wistful about lumberjacking and the long-gone days of "Hayward, Hurley, and Hell," but I had plenty to tide me over until the final jaw-dropping event of the summer: the Sawyer County Fair, held in late August just before the start of school.

Again, our car rolled toward Hayward, past the deer farm, where tourists fed and petted fenced-in fawns, past the log ranger station, partially hidden behind tall pine trees, past the Carnegie Library (which I would've killed to stop at), past the ancient Sawyer County Courthouse, and past the signs advertising the World Record Muskie in the Moccasin Bar. One year, Sam and I saw that muskie with Dad and Mom (teetotaling Grandma Bessie wasn't along), as Dad downed a beer and we all stared at the incredible volume of enclosed dioramas above and around the bar. Each glass box contained posed stuffed animals: skunks, raccoons, chipmunks, and rabbits fished, yodeled, gamboled, and ate. But, mostly, they boozed. (Did bars corrupt animals, as well as people, just as Grandma Bessie feared?) The dead deer, ducks, and fish on the walls could have entertained for hours, but the fair awaited.

After parking on the grassy area, Dad paid our admission at the fairground gate and we walked through the cow barn and

horse barn, where I wished I could be one of those 4-H kids bedded down in the box stall with my very own horse. We surveyed
the chickens, crowing roosters, and pigs with curly tails, pausing
a while longer to look at a strange animal I'd never seen before—a
guinea pig.

Shirley's 4-H project, showing sheep at the fair, included her
pet, Wilby. Shirley had bottle-fed the lamb, which was now grown
up. Despite the name, Wilby was a ewe. Jerry entered his Hampshire sheep in the Adult Open Class, and, like Wilby, many came
away with ribbons. We stopped by their pens to admire Wilby and
the prize buck and ewes. (Once, mischievous local lads opened
the fairground sheep pens late at night. Sheep scattered throughout town and caused a nightmare for Jerry, Millie, and others
as they rounded up the wayward flock, found munching on the
some of the finest lawns in Hayward.)

We continued past the rambling wooden grandstand and
harness race track nearby, then Dad and Sam scattered on their
own after agreeing to meet at the Methodist church food stand
for lunch. Lured by carnival barkers' coarse voices, Mom and I
wandered toward the midway and gaped at the spinning Ferris
wheel above, already bustling in midmorning.

But I led Mom straight to the pony rides. Tethered, tired
ponies trudged in a never-ending circle. Occasionally they were
allowed a brief rest, until the still-cheerful attendant swung up
into the saddles another crop of wide-eyed kids. The creak of the
leather saddle on my black and white spotted pony was music to
my cowboy ears. My pony obediently moved forward, pushed
ahead by his circle mates. When I was through, I hoped the ponies
would soon have a rest and a drink of water.

Then it was off toward the colorful carousel. Mom let me set
the pace, only hurrying me past the caramel apple that had caught
my eye and the clouds of pink and blue cotton candy—at least
until after dinner at noon. But I was further distracted from my

pursuit of more horses when I passed by a watery moat dancing with plastic pastel duckies. No one else was around, and the
barker knew he had me when we locked eyes. For a mere dime,
I could win! He waved his arm toward a dazzling row of stuffed
bears. I would lift the ducky revealing my prize-winning number. That was all. Guaranteed success. Mom stood quietly by as
I dropped my dime into his hand, selected my lucky ducky, and
waited for my fabulous prize.

Instead, I heard, "So sorry, little girl, but here's a nice prize."
He reached below, out of sight, and handed me a pitifully small
paper blowout toy. It had cost me my dime for the carousel ride.
Knowing I'd just learned one of life's hard lessons, Mom paid for
my ride up and down to cheery music on a horse of a different
kind, soothing my disappointment.

Then it was Mom's turn to be entertained, past the flower ar-
rangements, woven baskets, woodworking, handicrafts, sewing,
and 4-H projects to the clusters of ripe red tomatoes, golden potatoes, sets of perfect cucumbers, jars of beans and pickles, loaves
of homemade bread, layer cakes, cookies, and every sort of pie.
Although Mom never entered anything in the competitions, she
did murmur her approval, or disapproval, of the exhibits. The
non-air-conditioned building caused fluffy white frosting on
cakes to drip and run. The top of a three-layer cake, cemented
together with lemon filling, had slid off like a car in a ditch. Once-
fresh flowers sadly drooped and wilted in the heat, prompting
Mom to commiserate, "Oh, my, my . . ."

After supporting the local Methodists, who served a version
of potluck fare from a wooden shack with a drop-down counter,
we followed Sam around as he tried his luck. He threw darts at
balloons that moved just as the dart neared its target. He threw
baseballs at 78 records, to no avail. Broken plastic littered the
ground, implying that other big-prize winners had easily succeeded. But we suspected the barker tossed busted plastic on the

ground every day before opening his booth. Knocking over the stacked wooden milk bottles seemed easy enough, but neither Sam nor anyone else appeared to be carrying home the gigantic plush animals hanging from the tent. Sam was learning his hard life lessons, too.

Mom and I rode on the Ferris wheel overlooking the grounds where horses ran harness races. Each one-man driver sat low in his sulky and loudly lashed his long, light whip against the cart, *smack*, guiding his horse around the course. 'Round and 'round, the sulkies went. 'Round and 'round, the spinning Ferris wheel went. Whenever our seat stopped midair, swaying wildly, I felt slightly nauseous and was sure we'd be dumped to the ground below.

Before leaving the grounds, we all browsed through the commercial exhibits. Kids wore cardboard headbands with red feathers advertising some product or other. Adults wore fold-out cardboard sun visors touting water softeners. There were screen doors and windows, milk replacement powder for calves, home cleaning products, and a dasher-type (agitator) wringer washing machine that spun every color of plastic balls, instead of dirty clothes.

On our last pass through the midway, Sam and I shared a cone of sticky blue cotton candy, pulling off a bite that smelled sugary sweet and delicious . . . but disintegrated even as it reached our mouths. We walked past the deep-voiced barker inviting passers-by to swing the giant maul and ring the bell at the top—just as he did to demonstrate how very easy it was to win. I was sure Dad could do it, but he didn't even try. "Don't think we'd win that one, Peg," he said, wiser than I. I was beginning to think he was right.

But I knew I'd likely be back at the fair again next year, just to give the barkers another try. Hope springs eternal, especially in a young mind filled with a day of head-spinning delights.

Forget-Me-Not

I had never fully realized the extent of the impact my extended family had had upon my upbringing, my life being shaped to its core by the people and places in Mom's album of black-and-white photos, linking our moments in time. I have often mused about the quotation attributed to Dr. Seuss, "Sometimes you will never know the value of a moment until it becomes a memory." For me, some moments became priceless only well after the fact.

During the decade spanning the mid-1950s to the early 1960s, one by one, seven members of our immediate family suffered and died, culminating in head-spinning rounds of grief around 1960 when, in less than a two-year span, we would lose four.

Grandma Bessie was the first to go. Through the years, she had been cheered by her growing family, church community, and visitors from the neighborhood, and she delighted in keeping in touch with her far-flung relatives across the country. It was while preparing for visiting Indiana relatives that she fell; and it was then I pictured her broken leg stuffed into the backseat of Dad's car. I didn't know it meant the beginning of the end. She spent months in a wheelchair before passing away at the age of seventy-nine. Dad drove to the depot and sent a telegram to Indiana.

Though it was nearly the first of May, true to the North Country, nasty weather appeared in force on the day our Indiana relatives arrived for the funeral and continued on to the day of Grandma Bessie's burial: icy downpours, sloppy snow, frigid winds, and nearly impassable muddy roads. It took the biggest truck Exeland had to offer, expertly driven by problem-solver-bull-rescuer-gas-station-owner Harry Taylor. Traveling many extra miles, the long way around, he retrieved those traveling by train and forded them across flooded roads and overflowing bridges. The grieving Indiana guests shivered valiantly in inappropriate spring clothing and soaked, dainty lady shoes (suited best for Indiana's spring) while stalwartly huddling through the burial at the windswept cemetery to honor to the end Bessie's Meteor Hills spirit.

The summer I was eight and Sam was twelve, Dad built our addition; he didn't think it right that Grandpa Sam should be left alone any longer in the big house. Grandpa Sam would move into one of the new bedrooms and Sam and I would share the other. One day, Dad drove the tractor and hay wagon to Grandpa Sam's house and returned later that afternoon with his big double bed; a tall, old-fashioned dresser with a mirror that reached all the way to the ceiling; a trunk full of diaries, photos, and letters; an ancient Boston rocker that had been in the family for decades; and a table-top wooden clock with a pendulum that *donged* once on the half-hour and chimed every hour. It took me weeks to sleep through the twelve chimes at midnight.

As Grandpa Sam rocked, he told stories about his youth, people he'd known, places he'd been, and family history. How I wish I'd listened better.

∽

One day around two years later, Mom told me and Sam, "Better get your chores done early. We're going to Grandma Walhovd's." Going to Grandma's house on a school day? That seemed very odd. We never visited, except on Sunday. I wondered why but didn't think too much about it; after finishing my chores, I was too busy reading a book on my current favorite topic, pioneers. Mom and Dad hurriedly finished supper and milking earlier than usual. Soon we were off, me with nose in book. At Grandpa and Grandma Walhovd's house, the adults seemed engaged in serious conversation in the next room. Still engrossed in reading, I paid little attention. Later, I realized that we had gone so my parents could tell her parents that Mom had been diagnosed with cancer.

Cancer. The word everyone feared. The word that would color the remainder of my childhood. The word that changed my ten-year-old life. Later that spring, on an otherwise ordinary day in fifth grade on the Exeland school playground, I overheard Ann whisper to her friend Jenny, "Peggy's mom's got cancer."

The dreadful words hanging in the air stabbed through the fabric of my world, cut like a knife, and slashed the routine of the childhood I knew. I hadn't told anyone; my best friends, Richard, Verna, and Margie knew Mom was sick, but I hadn't even told them. Mom's illness had been quietly shared within each neighbor's family, as a support for us. How did Ann know? She lived in Exeland. Her words turned our family business into nothing more than small-town gossip. If a secret was to be shared, wasn't it mine to share? My world changed even more that day when I knew I was the object of classroom whispers.

I told no one of that incident at school, not my best friends, not Mom and Dad, not my brother. No one. Had a school counselor been available in the 1950s, it is unlikely I would have told him or her either. Instead, I watched, listened, and thought things over in my own time and in my own way.

Soon, Mom underwent surgery and spent the first of many days at St. Mary's Hospital in Ladysmith. It was on that sunny day, when mayflowers bloomed in the woods, that I first realized Richard's mother, Ruth, would be there for me. Although she used few words, whether from shyness or because she found them unnecessary, her message was clear from her actions. She knew to simply be there. I spent that day playing with her kids. When Dad returned from the hospital, he spoke quietly with Ruth. She took over in her own unassuming way, reassuring Dad, and she knew just what to do for me, too. That day she had watched us play with a batch of young farm kittens. As Dad started up the car, Ruth tucked a lively striped kitten into my arms for the ride to our house. Tabby became my constant companion through the rest of childhood, carrying me into and through adolescence, giving solace during the years of multiple family deaths.

Not even two years after Mom's diagnosis and surgery, Grandpa Walhovd suffered a stroke and passed away in February. Grandma Walhovd never even knew of her husband's death; earlier that same month, she had gone to a nursing home with severe sudden dementia. She died less than two months later, in early April. Uncle Vic passed away that spring, too. Then Grandpa Sam fell and broke his hip after having "taken a stroll around" outside. He died at the end of that April.

Four funerals in three months.

Grandpa Sam's creaky Boston rocker went to the junk heap, same as the ancient trunks, though we did keep piles of his old photographs and some letters between Grandpa Sam and Grandma Bessie. From Grandma and Grandpa Walhovd, Mom got her gas kitchen stove and our first black-and-white TV. And another table-top chiming clock. The two clocks weren't synchronized; it took me weeks, again, to sleep through the now twenty-four midnight chimes.

That summer, we received more stunning news: Jerry, only in his late fifties, had suffered a debilitating stroke. The hard feelings over the school consolidation were put aside, and our families came together again for mutual support. Doctors operated and removed Jerry's large brain tumor, and he regained part of his ability to move, function, and speak, although haltingly. But by the end of December 1960, he was gone, too.

Grandma Bessie and Grandpa Sam gone. Grandpa and Grandma Walhovd gone. Vic and Jerry gone. And Mom's health was deteriorating, too. Compounding the troubles, instead of the prosperity we'd expected, we were now in debt.

Like most farmers, we had no health insurance. Dad logged more in the winters to pay for mounting bills, added a few more beef cattle he could sell for extra cash, and kept on with the maple syrup production and sales, as best he could. Then he went to Edith at the bank, to take out a mortgage on the farm. I was unaware of our financial straits until one evening at the kitchen table when I watched Dad open an envelope of money collected from our church. Donations. For us. Finally, Dad reluctantly put Grandpa Sam and Grandma Bessie's vacant house up for sale.

Over the next couple of years throughout the cruelty of illness, repeated kind words, deeds, and visits kept us going. Ruth stood watchfully by. I spent countless hours, days, and evenings with her family, going to the Sawyer County Fair, Spooner Rodeo, and Lumberjack World Championships in Hayward and swimming at Windfall Lake. Ruth gave me normalcy when the threatening cloud of deaths and illness hung overhead. There must have been times when she longed to gather her family around her without the kid from down the road, but if she did, I never knew it. And Tabby helped me through it all, just as Ruth knew he would.

As Mom grew weaker and her fatigue increased, her ability to be active in the household became more limited. She hung up her apron; I learned to cook. Sam and I cleaned the house. When a neighbor asked, as a compliment, "How do you keep your house so neat and clean?" I quipped, "Well, guess we just don't have time to get it dirty."

Mom no longer sang. The piano fell silent.

From Sears or some other catalog, Mom mail-ordered a set of oil paints to help her pass the time. Although Sam and I no longer raced our bikes to the mailbox, on the day of the paints' expected arrival, I pedaled furiously to and from the mailbox, returning with the colorful box and pussy willows in my basket. With a little practice, Mom's artistic bent was reawakened. She painted a handful of canvases, including a landscape or two for our wall. That's when Rob gave Mom the picture of sandhill cranes that he had ripped out of one of his *Wisconsin Conservation Bulletin* publications. Would she paint it for him—as a memento from his youth, when vast numbers of cranes flew Wisconsin skies? She began to paint an eight- by ten-inch scene of sandhill cranes on a marsh. But one day, Mom laid down her brush and didn't pick it up again; she no longer joked about being the next Grandma Moses. Amidst the chaos of illness, Dad stored the unfinished paintings, including Rob's, in the attic.

Despite hospitalizations for rigorous treatments, the illness had progressed. Dr. Pagel gave Mom sample drugs she would not have to pay for, and a local nurse drove to our house to give Mom the shots she needed. Mom moved to a warmer bedroom in our house.

Under the stress of so much family trauma, Dad suffered a stroke. Dr. Pagel said he must rest. Sam and I did the milking; neighbors came to help with chores and finish putting up the hay crop. The next winter, Dad was able to go back to work. Then a logging chain broke, landed a blow to Dad's face, and jeopardized the sight in his eye.

Slowly, I became aware that there was every possibility I might lose both of my parents.

A community of kindnesses kept us going. The school lunch lady, Grandma Fairman, brought over a special burnt-sugar layer cake served on a red and white plate; Aunt Millie brought baked rice pudding, pies, and loaves of bread; Ole brought ice cream and cookies. Inee invited us for dinner while Mom rested at home. Uncharacteristically, she didn't pull out every pan from her cupboard for the two of us to wash. I watched Inee's silent tears drip into her dishwater as we stood side by side at her kitchen sink.

Neighbors Mom had walked to visit on our Sunday sojourns now came to visit her. Lena Erickson came, her rich voice chattering on about this and that, the latest news about her flocks of geese, ducks, turkeys, and chickens, before she finally prayed over Mom. This time, I appreciated it.

Nola, Flossie, Bernice, Ethel, and Elaine, from the local Homemakers Club, brought cheery news of the neighborhood and tomato-hamburger-macaroni hot dishes. Christmas 1961, the Ladies Aid showered Mom with a basket of small gifts.

Ole, Rob, and Inee came often that last winter. When Ole came to visit, he no longer greeted us with a joke at his arrival or asked as he so often had, "Where's th' ol' man?" (referring to Dad's being four months older). Their conversations became very quiet and subdued. Mom's lungs couldn't stand Rob's pipe smoke. Dad asked Rob not to light up anymore while he and Inee visited.

Mom had read Norman Vincent Peale's 1952 book, *The Power of Positive Thinking*. Through it all Mom remained outwardly optimistic. Her sentences often began with, "When I get well. . . ." But no one needed to tell me Mom was getting worse that last winter, spring, summer, and early fall. After returning from one of many trips to the hospital, Dad told me one night, "This time, Mom likely isn't coming home."

The next day, Dad, Sam, and I drove to Ladysmith. I stood at Mom's bedside as she looked wearily at a bed of blood red cannas in full fall color below the hospital window. "Next year," she murmured, "we'll plant flowers."

Overnight, right on schedule for northern Wisconsin, the first hard-killing frost fell and drained all color from growing things.

That day, September 20, 1962, Mom died; the cannas sagged in stark drooping black. Inside, I did, too.

Dad was at the hospital already; Sam drove me there. News traveled fast. From the hospital window, I watched Rob and Inee slowly inch the Green Hornet into a parking spot below. Aunts Millie and Bert arrived a short time later. Dad called Mom's siblings on the hospital phone and notified distant relatives. That afternoon, Dad, Sam, and I returned home in silence. We dreaded entering our home without Mom. But there we found a warm, welcoming group of neighbor ladies and a hot meal already on the table in our kitchen.

On the day of Mom's funeral, I scanned the church looking for, and finding, Verna, Margie, Richard, and their parents, especially Ruth. The church filled with neighbors, relatives, friends, and flowers. Bud, my school bus driver, and the kids on bus number eight had collected money to send flowers. Basement tables were covered with food. Ma Carter's Cafe sent homemade pies. The Indiana relatives gathered. Again. This time on a stunningly beautiful fall day.

For Mom.

The funeral procession moved slowly from the church onto Main Street, Exeland. I looked out the window and noticed that even though I was only from the town, Exeland folks *of* the town had stopped to pause in silence as Mom's hearse slowly drove on its way to Windfall Cemetery.

After Mom's death, Dad worked to pay off the medical bills and the mortgage on the farm and saved money to send me to college. He spent time enjoying the natural world and remained on Badger Creek, though the rest of his world had drastically changed with the loss of Mom, family members, Sam leaving the farm for work elsewhere, and my going off to college.

On a weekend visit, I wandered to the back forty for a survey, as I always did, to especially admire the two soaring pine trees on the edge of the woods along the western horizon. Old Joe, head down, grazed contentedly in the pasture, just as always. To my dismay, the pines were gone.

"Dad!" I blurted once back in the kitchen, "What happened to the pine trees behind the barn?"

He poured another mug of his lumberjack coffee, handed it to me, and looked long before replying, "Well, Peg, I cut those pines to pay this year's tuition." Dad's use of timber as currency was forever driven home to me.

Old Joe and Dad retired together from the bulk of farm chores, but the two of them continued to skid logs and cut wood in the wintertime, long after tractors took over other chores. Dad seemed to have a special bond with Joe, too. Perhaps that's why Dad had kept him for so long; he had gradually sold off the other animals: first the milk cows, then the beef animals, and finally, even Queenie. Now, Joe roamed the back pasture alone, and Dad walked out to check on him every day or so and fed him hay and grain in his winter shelter.

"Where's Joe?" I asked one day when I was home for another visit. Dad stood quiet for a moment, then told me that he'd found Joe tangled in a fence. As farmers must sometimes do, Dad had to dispatch his old companion. He never spoke of it again, and I did not have the heart to ask for details.

Many years after Joe's demise, I wandered across one of the fields that Joe had mowed, cultivated, plowed, and traversed so many times. I sat down to meditate on a wooded rock pile overlooking Badger Creek. While remembering old times, I noticed an odd-shaped object on the rocks. It took a while to bring into focus the worn and weather-bleached form of a horse's skull. The long bones could only be those of Joe. Dad must have moved his remains to the shelter of the woods for Joe's final resting place. Old Joe's head was still "bending low."

When he was in his eighties, Dad still harvested his own wood to heat the house as he always had. And he cut some to create. He spent many happy hours making tables, benches, chairs, and bookshelves for himself and others. The Exeland Historical Society asked Dad to build a large wooden cross to be placed on the banks of the Chippewa River at the Sawyer County historical site marking where the first white settler, Charlie Bellile, had arrived as early as the 1840s. A plaque at the site also notes an early Catholic mission church and cemetery at Bellile, near Exeland. Dad took me to show off his handiwork.

As he and I looked across the Chippewa, I heard a strange, trumpeting sound: sandhill cranes were calling as they flew in a circle around a field before landing—the first time I had ever seen cranes in northern Wisconsin. In a remarkable comeback, sandhill cranes now flock in huge numbers in every part of the state, along with a small but growing population of whooping cranes—scenes Rob and Mom never could have imagined.

A few years later, Dad was moving from the home he'd built and spent nearly his entire adult life in. I watched new owners carry the forest green canoe away, resort owners, I realized too late, who didn't know anything about Dad's handmade, time-honored traditional craft of stretched canvas and wood. I shuddered as I pictured her classic, still-beautiful-though-brittle form teetering atop a boulder on the Chippewa River. Or skewered on a rock, her tender ribs smashed, and a gaping hole in her side. Still, the river would be familiar territory for her. And Dad was well beyond his canoeing days. But as we sorted more of his life, I felt that in haste I'd foolishly suggested Dad's cherished possession be sold.

Dad examined his ancient things as we packed them away, pitched them, or organized them for the auction: the manual for the 1940s Case tractor; a double-bit axe and a crosscut saw from Dad's lumberjack days; Joe's and Queenie's leather harnesses;

the three-legged milking stool and kerosene lantern; still sturdy Malone pants and buffalo check wool hats with flaps; Belle's blue porcelain teapot; Mom's gingham apron decorated with Inee's cross stitch; multitudes of blue Mason two-quart canning jars; the maple syrup hydrometer; Dad's wooden muskie baits, carved by Rob, and which, Dad pointed out, had embedded teeth marks from the muskies that got away. The woolen Canadian blanket, Dad's snowshoes, and my handmade skis. Dad lingered as he picked up the Indian trade axe he'd found while clearing our land.

Dad really needed a new table, but it didn't seem right to leave the old kitchen table behind. So we didn't. Although she could barely stand on her old legs, unless propped tightly against the kitchen wall for support, she came away with Dad. He cluttered her surface, keeping everything he needed within easy reach of his chair. When friends visited, he cleared her off, "boiled the hell outta" some coffee, and put together a pretty good meal, cheering all. Yet, taken away from the home they'd known for so long, Dad and the kitchen table faded quickly.

One brilliant and crisp day, I took Dad for a last autumn ride through our old neighborhood. As we passed the creek, he said, "More 'n anything, Peg, I miss being able to take a walk along Badger Creek."

Dad passed away on April 6, 1996, and was buried on April 9, his eighty-seventh birthday. His obituary included the sentiment he often repeated: "I worked the land, and the land was good to me."

Epilogue

Grandma Bessie's humor was reserved, but if she knew the trick her aged homestead garden played on me, she would have laughed heartily. I once dug thick clumps of her stunted, overgrown daffodil bulbs to transplant in my own garden, over two hundred miles away. Among them was one bulb I could not identify. The daffodils bloomed profusely, and I waited patiently for the perfectly healthy mystery bulb to thrive. But year after year it failed to produce a flower, only wide leaves, which dried, turned yellow, and faded away. It became overtaken by other plants and I concluded there was no hope. Then one early August day, amid the growth of companion plants, I spotted it: a tall green stalk topped by a pale lavender lily.

"Why, that's a Naked Lady!" my neighbor shrieked in delight.

My prim grandmother would have been shocked to hear her flower thus described, deeming "Naked Lady" improper. No doubt Grandma Bessie knew it by its other name: Surprise Lily.

Some gardeners, like me, are avid watchers of the plant world and find hope in searching for flora in the garden patches of old homesteads that are abandoned of the living yet still ripe with memories. Over the years, I have made many pilgrimages to Grandpa Sam and Grandma Bessie's old place, at first with my father. Years after the barn was gone, Dad and I took buckets to the site, buckets not long empty. Alongside the remains of the doorway, a once-giant pile of horse manure had crumbled into dirt. We unearthed the soil camouflaged under a veil of thriving weeds; beneath the crusted weed bed lay treasure. A planter's gift: black topsoil. From ashes to ashes, dust to dust, new life; gardeners find healing by digging in the dirt.

Now that Dad is gone, I tramp the tote road woods alone, my nose to the ground in search of spring, breathing in air filled with promise. Sniffing out the fragrant woodland phlox, I find the native blue flowers on the sandy hillside, just as they were when Sam and Bessie Prilaman arrived in the spring of 1916. The Prilaman northern Wisconsin story began then, on this hillside in the Meteor Hills. They settled their land, taking it from cutover wilderness to tiny farm. Just enough. Now, even as the land returns to its origins, it bears their mark: there are remnants of the asparagus patch, the apple orchard, and the aromatic black poplar that Grandpa Sam called his prized "balm of Gilead."

And, still, I step through Grandma Bessie's patches of mint-scented bee balm and periwinkle—vinca vine—whose glossy leaves remain green even under winter's snow. She planted it just outside their cabin door, and now the tough trailing plant is ever-expanding, covering the hillside. In May and June, its blue flowers match the sky.

Like vinca vine, gardens weave together a long trail of family history. Visions of past gardens remind me of the time when, in their boyhood, Dad and Jerry had gathered the rich brown soil in the fall before it froze solid. Grandma Bessie then stored the earth in homemade crates in the cellar—come March, to be sown with last season's dried tomato seeds, which she saved from only the most worthy, the biggest and juiciest, tomatoes. The young plants gathered warmth in her sunny east window overlooking Badger Creek. Lengthening rays gave those leggy stalks strength until Grandma Bessie could safely transplant them when frost no longer threatened. By fall, the bountiful results rested in gleaming jars: Grandma Bessie's ruby jewels. Her winter table would be loaded with the garden's gifts after many a spring, summer, and fall walk down the garden path. I can see her now: Grandma Bessie wearing her wide-brimmed sunbonnet for an afternoon spent tending the garden.

Gardeners' lives are measured in seasons and in growth. They mark milestones: the white narcissus in May that bloomed on my parents' wedding day in 1938, the pastel-colored gladiolas that overflowed in Grandma Bessie's arms on her fiftieth wedding anniversary in 1947, the prized maroon peony that we planted on Grandma Bessie's grave in 1954. It blooms each June, sweet and simple.

My grandparents, like all gardeners, were happy with small successes; they would be pleased to know that to this day their daffodils still bloom every April. Grandma Bessie planted the hardy bulbs one hundred years ago, and those beaming buttery flowers still remain the earliest omen of spring as they make their debut in splashes across the landscape, announcing that the long-awaited season has begun. Winter is over.

～

At the start of the school year in 1998, my teaching career skidded to a halt. I was fifty-two when Dr. Updike told me he had not seen a case of primary amyloidosis in his forty-three years of practice. But he knew what it was, and he knew the only treatment. He explained the auto-immune blood disease produces an unusable protein that shuts down organs: kidneys, in my case. At that time, it was rare and uniformly fatal. My only hope would be a stem cell transplant at the University of Wisconsin Hospital—the first such transplant done at UW for this disease. That November, the long ordeal began when doctors harvested my stem cells and administered ultra-high chemotherapy in preparation.

On the first day of spring in 1999, I entered a sealed isolation unit. Medical students peered into my charts, monitors clicked and clanged, my throat blistered, and I saw things that were not there. My life hovered in the balance, impossibly grim. Yet, in my mind, Badger Creek's water trickled soothingly over stones, and sunshine filtered warmly through the leafy glades to the forest floor. In moments of lucidity, I wondered if I would survive and ever be well enough to sit again on the banks of Badger Creek.

Cheers erupted when the first new cell appeared and launched the beginning of my hundred-day journey toward regaining healthy cells. I had begun with November's snow piled outside my hospital window; weakly turning my head now, I was astonished to see bright yellow daffodils in bloom. Real daffodils, not hallucinated. Daffodils like Grandma Bessie's. And in time, I did return to health and to Badger Creek.

In 2000, my pilgrimage took place alone, in grateful appreciation. From Grandma Bessie and Grandpa Sam's homestead site, I followed Badger Creek upstream on the tote road through the woods, toward the series of small drops locally known as the Seven Steps. I ambled past gurgling springs and stopped to rest on a moss-covered boulder, surveying the rose-colored wall of sheer rock and the pool below it, near where Mom and Dad had once watered weary horses while they built their log home. As I sat pondering life, I remembered the day I had spotted something spinning in the frothy creek water. What I had at first taken to be one of my errant baseballs turned out to be a perfectly round rock, as mysterious to me then as a child as Mom's death would be— and as my illness and recovery were now. Mysteries—remaining part of the scheme of living.

Walking along the banks of Badger, I looked for trout lurking in the stream. On the carpeted forest floor, I spied wild ginger, fading trout lilies, and jack-in-the-pulpits and noticed the trilliums had turned from white to pale purple. While breathing in

Mom's forget-me-nots on the creek

fresh air, feeling cool breezes on my skin, and listening to Badger's lulling sounds, I continued to review the scenes of my childhood, savoring memories. Little had changed—an occasional windstorm, a lightning strike, the cutting of the hardwood. But still the creek flowed on, shifting, yet the same.

Around the stream's horseshoe bend and just short of Dad's favorite trout hangout, I saw a patch of pale robin's egg blue, surrounded by hints of green. Tiny plants had taken root: a sea of forget-me-nots had spread across the gravel and rocks, and solid lines of blue covered the sandy creek bank. Once upon a time, Mom had planted them to brighten our dooryard. The delicate yet resilient flowers remain a fitting legacy to her and Grandma Walhovd, from whom Mom received them.

My mother, Vera Walhovd Prilaman, died when I was fourteen and she was forty-nine. Her life ended much too soon. But once, she planted forget-me-nots. And I am home again, on Badger Creek.

Mom on Christmas Eve, 1956

Acknowledgments

Like most people, written material has been part of my life since I can remember. Dad spent hours at the kitchen table, book or magazine over his crossed knee; Mom did her share of reading too (teacup in one hand), including reading to me.

My childhood was also influenced by my maternal grandmother's history-collecting neighbor, Mrs. Ethel Chappelle. She ignited my interest in the history of the area with her book, *Around the Four Corners*. Her example sparked a desire to tell the story of saved letters, papers, and photos left behind by my mother and my grandparents.

My father's side of the family contributed to encouraging me to write, as well: Gertrude Finney, my paternal grandmother's niece, wrote historical fiction for juveniles based on 1860s family history (although editors softened the grim homesteading story). *The Plums Hang High* was published in 1955 and widely read (could authorship be in my bloodline?).

The Farm on Badger Creek: Memories of a Midwest Girlhood began with stories I wrote over a span of twenty years, at first just for myself, then discrete sketches written to read aloud to a Reminiscence Writing class I joined in 2003. We—mostly women ages sixty to ninety-five—wrote to keep the stories of our lives alive for future generations. And to remember, to laugh, and to cry—together. Along with our first facilitator, Ann Short, and our facilitator for the past twelve years, Katherine Perreth, my classmates urged me to someday share my stories. Growing up in northern Wisconsin in the 1950s was quite different from most class members' backgrounds. Someday, I thought. Maybe . . . someday.

In the fall of 2016, I decided *someday* had arrived. It was time to take a break from the swirling negative news cycle encompassing the nation. Time to pay tribute to the family and community I remembered during a kinder time. Time to revisit the values of country life, even as the memories of that way of life fade away.

Thank you to many friends who offered words of encouragement. To Duane, my husband, who offered support, gave technical advice, and kept the computer running and printing.

Thank you to Kate Thompson at the Wisconsin Historical Society Press for publishing this story of a 1950s girlhood in northwestern Wisconsin and to my editor, Rachel Cordasco, who so competently and kindly blew away the chaff, winnowing out this story of our community.

I am grateful and give credit to my first editor, Katherine Perreth, of Reminiscence Writers, who with good humor listened, questioned, encouraged, provided advice, and skillfully helped me weave together the time, people, and personality of my 1950s community.

I write to honor those pioneers of old, my family and wider community, as our lives unfolded near the banks of Badger Creek. And, finally, I write in tribute to my ancestral home in the Meteor Hills on the occasion of the area's centennial celebration: Town of Meteor, 1919–2019.

Happy Birthday, Meteor!

About the Author

Peggy Prilaman Marxen grew up on a farm in the hills of northwestern Wisconsin in Sawyer County. Her parents and teachers instilled a love of reading and writing in her at an early age, a love that she continued to cultivate throughout her rewarding three decades of teaching.

Fueled by pleasant memories of Valley View School and the 1950s notion that girls could only become nurses, teachers, or secretaries, Peggy decided to work in education. Her father insisted that she aim to go to college but also prepare for alternative work. Thus, her high school college prep courses also included Mr. Kaiser's excellent secretarial curriculum.

Peggy retired from the Middleton–Cross Plains Area School District after teaching there for thirty-three years. She and her husband, Duane, live in Middleton, Wisconsin, with occasional escapes to Exeland and Badger Creek.